BENJAMIN PEIRCE:

"Father of Pure Mathematics" in America

BENJAMIN PEIRCE:

"Father of Pure Mathematics" in America

I. Bernard Cohen, editor

ARNO PRESS

A New York Times Company
New York • 1980

Publisher's Note: This book has been reproduced from the best available copy.

Editorial Supervision: Steve Bedney

———————

Reprint Edition 1980 by Arno Press Inc.

Copyright © 1980 by Arno Press Inc.

"Benjamin Peirce's Linear Associative Algebra and C.S. Peirce,"
by Raymond Clare Archibald, originally published in the
American Mathematical Monthly, vol. 34, 1927, has
been reprinted by permission of the Mathematical Association
of America

"BENJAMIN PEIRCE: Mathematician and Philosopher" has
been reprinted with the permission of the *Journal of
the History of Ideas*, which published Sven R. Peterson's
article in vol. 16 (1955), pp. 89-112

Linear Associative Algebra has been reprinted from a copy in
the Bowdoin College Library

THREE CENTURIES OF SCIENCE IN AMERICA
ISBN for complete set: 0-405-12525-9
See last pages of this volume for titles.

Manufactured in the United States of America

———————

Library of Congress Cataloging in Publication Data

Main entry under title:

Benjamin Peirce : "father of pure mathematics"
    in America.

    (Three centuries of science in America)
    Bibliography: p.
    CONTENTS: Archibald, R. C. Benjamin Peirce, 1925.
--Archibald, R. C. Benjamin Peirce's linear
associative algebra and C. S. Peirce. 1927. [etc.]
    1. Algebras, Linear--Addresses, essays, lectures.
2. Peirce, Benjamin, 1809-1880--Addresses, essays,
lectures. 3. Mathematicians--United States--
Addresses, essays, lectures. I. Cohen,
I. Bernard, 1914-   II. Series.
QA184.B44   512'.5   79-7981
ISBN 0-405-12563-1

CONTENTS

BENJAMIN PEIRCE:
1809-1880

Raymond Clare Archibald

BENJAMIN PEIRCE

1809–1880

BIOGRAPHICAL SKETCH AND BIBLIOGRAPHY

BY

Professor Raymond Clare Archibald
Brown University, Providence, R. I.

REMINISCENCES

BY

President Emeritus Charles W. Eliot
President A. Lawrence Lowell
Professor Emeritus W. E. Byerly
Harvard University

Chancellor Arnold B. Chace
Brown University

OBERLIN, OHIO
The Mathematical Association of America
1925

FOREWORD

This monograph was originally published in the *American Mathematical Monthly*, January, 1925. The Mathematical Association of America issues it in separate form, with four new portraits and additional notes, so as to meet the needs of libraries and individuals throughout the world, interested in the outstanding figure in American Mathematics during the first three quarters of the nineteenth century.

CONTENTS

BENJAMIN PEIRCE, 1845 (?)

Earliest known portrait from a hitherto unpublished daguerreotype in the possession of his grandson, Benjamin P. Ellis.

BENJAMIN PEIRCE, 1857

From a hitherto unpublished painting by Daniel Huntington. Reproduced through the courtesy of George A. Plimpton of New York City, who has recently presented the portrait to Harvard University.

BENJAMIN PEIRCE

I. Reminiscences of Peirce.

By President Emeritus Charles W. Eliot,[1] Harvard University.

Benjamin Peirce graduated at Harvard College with the degree of A.B. in 1829. Two years later he was appointed Tutor and in 1833 University Professor of Mathematics and Natural Philosophy. This was an unendowed professorship; and its creation was one of President Quincy's enterprising adventures in the enlargement of Harvard's teaching staff. The President was doubtless

---

[1] The authors of the "Reminiscences" of Benjamin Peirce, presented herewith, were all his former students and each has done something notable in mathematics. President Emeritus Eliot was a student during 1849–53. He was a tutor of mathematics in Harvard College 1854–58, and assistant professor of mathematics 1858–61; he was also assistant professor of chemistry 1858–63. James Mills Peirce, son of Benjamin, and classmate of Eliot, was appointed tutor of mathematics at the same time. In his *Analytic Geometry*, published in 1857, Tutor Peirce acknowledged that "whatever merit the book may have is owing, in a great degree, to the assistance of Mr. C. W. Eliot." President Eliot has described these early years as follows (*Report of the Harvard Class of 1853*, Cambridge, 1913, p. 98): "Tutor Peirce chose the Freshman class, leaving me the Sophomore class in that year [1854–55]. After a year's experience, we applied some new recitation-room methods which made the mathematical instruction more effective. Finding the existing method of conducting oral examinations twice a year in the presence of visiting committees of the Board of Overseers very unsatisfactory as a test of the students' knowledge and capacity, we asked leave of the Faculty to conduct the mathematical examinations of the Freshmen and Sophomores in writing. After a good deal of hesitation the Faculty granted us leave to make the experiment; and these examinations were the first examinations in writing ever conducted for entire classes in Harvard College. The innovation was gradually adopted in other departments, and ultimately spread to the whole University.

"I tried to make the teaching of mathematics to the Freshmen and Sophomores as concrete as possible, and to illustrate its principles with practical applications. For example, while the class was studying trigonometry, I taught simple surveying to a group of volunteers, and with their help made a survey of the streets and open spaces of that part of Cambridge which lies within a mile and a half of University Hall. These volunteers made under my direction a careful map of what was then the College Yard, with every building, path, and tree delineated thereon—a map which is preserved in the college library."

President Lowell was a student under Peirce 1873–77, and his paper on "Surfaces of the second order as treated by quaternions," read before the American Academy of Arts and Sciences, was published in its *Proceedings* (vol. 13, 1878, pp. 222–250).

Professor Byerly was a student under Peirce 1867–71. He was assistant professor of mathematics at Cornell University 1873–76; assistant professor of mathematics at Harvard College 1876–81; professor 1881–1906; Perkins professor of mathematics 1906–1913. Since 1913 he has been Perkins professor emeritus. He was the first one (in 1873) to receive the degree of Doctor of Philosophy, in mathematics, at Harvard University; his thesis was entitled "The heat of the sun." He is the author of mathematical articles, pamphlets, and textbooks.

Chancellor Chace studied with Peirce in 1878–79, and his paper on "A certain class of cubic surfaces treated by quaternions," published in the *American Journal of Mathematics* (vol. 2, 1879, pp. 315–323), was a result. The Chancellor has recently prepared a translation, with commentary and notes, of the Rhind mathematical papyrus, which is about to be sent to the press.

R. C. Archibald.

supported in this adventure by Nathaniel Bowditch, who was then at the height of his influence as a Fellow of the Corporation (the President and Fellows of Harvard College). As soon as the endowed Perkins Professorship of Astronomy and Mathematics was established (1842), Benjamin Peirce was transferred to that chair, which he held till his death in 1880.

Benjamin Peirce was never a professor of Mathematics only. In the title of the University professorship he held, the broad subject of Natural Philosophy appeared, and in the title of the Perkins professorship Astronomy was the first subject named. These titles represented the real breadth of Benjamin Peirce's mental interests and imaginative powers, and this breadth characterized his teaching in Harvard College from beginning to end.

He was no teacher in the ordinary sense of that word. His method was that of the lecture or monologue, his students never being invited to become active themselves in the lecture room. He would stand on a platform raised two steps above the floor of the room, and chalk in hand cover the slates which filled the whole side of the room with figures, as he slowly passed along the platform; but his scanty talk was hardly addressed to the students who sat below trying to take notes of what he said and wrote on the slates. No question ever went out to the class, the majority of whom apprehended imperfectly what Professor Peirce was saying.

When I entered College in 1849 Professor Peirce had ceased to have to do with the elementary courses in Mathematics. He taught only students who had been through the two years of prescribed Mathematics and had elected to attend his courses, which were given three times a week throughout the junior and senior years. Two or three times in the course of the hour, Professor Peirce would stop for a moment or two to give opportunity for the members of the class to ask questions or seek explanations; and these opportunities were utilized by all the members who really wanted to learn. If a question interested him, he would praise the questioner, and answer it in a way, giving his own interpretation to the question. If he did not like the form of the student's question, or the manner in which it was asked, he would not answer it at all, but sometimes would address an admonition to the student himself which went home.

One day in my senior year, when Professor Peirce had already acquired the habit of giving me the highest possible marks on all my notes of his lectures and on every other exercise for which marks could be given, to the great concern of my competitor for the first place in the class, a concern which he liked to communicate to me his next door neighbor in Hollis—I graduated second—I ventured to say that what he had just been saying to us about functions and infinitesimal variables seemed to me to be theories or imaginations rather than facts or realities. Professor Peirce looked at me gravely, and remarked gently, "Eliot, your trouble is that your mind has a skeptical turn. Be on your guard against that tendency or it will hurt your career." That was new light to me; for I had never thought at all about my own turns of mind. The diagnosis was correct.

In spite of the defects of his method of teaching, Benjamin Peirce was a very inspiring and stimulating teacher. He dealt with great subjects and pursued abstract themes before his students in a way they could not grasp or follow, but nevertheless filled them with admiration and reverence. His example was much more than his word. I remember that this great master began one day with unusual promptness to put on the slates a series of calculations and formulæ in which he seemed to be much interested. He said but little; but wrote diligently with the chalk, stopping now and then to examine his work and to rub out some of it, but only to resume it, and go on eagerly. The class before him said not a word, took notes as well as they could of what he wrote on the slates, and watched him. Suddenly near the end of the hour the worker looked despairingly at the contents of the last slate he had filled, turned to the class, and remarked, "there is an error somewhere in this work, but I cannot see where it is. This last line— the conclusion—is obviously wrong." Whereupon he seized the rubber and rapidly rubbed out everything he had put on the slates. Professor Peirce sat down in his armchair visibly fatigued. The class slowly folded their notebooks and departed without a word, even to each other. I, for one, have always remembered vividly that hour's spectacle.

In 1862, Thomas Hill was elected President of Harvard University. One of his first measures was the institution of courses of lectures open to graduate and other advanced students, and called University Lectures. President Hill's idea was to give advanced students living in Cambridge and the interested Cambridge public opportunities to hear the best scholars and scientists of the country speaking on their favorite subjects, the subjects in which they had won distinction or renown. Benjamin Peirce was one of the first persons to be appointed University lecturer, and he served gladly in this capacity in five different years. The University Lectures were not to be technical, though advanced. They were to be stimulating as well as informing, and women were encouraged to attend them as well as men. Benjamin Peirce's lectures dealt, to be sure, with the higher mathematics, but also with theories of the universe and the infinities in nature, and with man's power to deal with infinities and infinitesimals alike. His University Lectures were many a time way over the heads of his audience, but his aspect, his manner, and his whole personality held and delighted them. An intelligent Cambridge matron who had just come home from one of Professor Peirce's lectures was asked by her wondering family what she had got out of the lecture. "I could not understand much that he said; but it was *splendid*. The only thing I now remember in the whole lecture is this—'Incline the mind to an angle of 45°, and periodicity becomes non-periodicity and the ideal becomes real.'"

When Professor Bache retired from the superintendency of the U. S. Coast Survey, he procured the appointment of his intimate friend Benjamin Peirce as his successor in the superintendency. Those of us who had long known Professor Peirce heard of this action with amazement. We had never supposed that he had any business faculty whatever, or any liking for administrative work. A very important part of the Superintendent's function was to procure from

Committees of Congress appropriations adequate to support the varied activities of the Survey on sea and land. Within a few months it appeared that Benjamin Peirce persuaded Congressmen and Congressional Committees to vote much more money to the Coast Survey than they had ever voted before. This was a legitimate effect of Benjamin Peirce's personality, of his aspect, his speech, his obvious disinterestedness, and his conviction that the true greatness of nations grew out of their fostering of education, science, and art.

In his younger days Benjamin Peirce enjoyed taking part in private theatricals. As an actor he was apt to be too violent and impetuous; but he was always interesting. He had, indeed, a gift for dramatic expression which served him well in many incidents, both comical and tragical, of his maturer life. For this reason, among others, only persons who saw and heard him can fully appreciate the influence of his life and work.

II. REMINISCENCES.

By President A. LAWRENCE LOWELL, Harvard University.

Looking back over the space of fifty years since I entered Harvard College, Benjamin Peirce still impresses me as having the most massive intellect with which I have ever come into close contact, and as being the most profoundly inspiring teacher that I ever had. His personal appearance, his powerful frame, and his majestic head seemed in harmony with his brain.

The amount of instruction in mathematics then given was small compared with what is offered in any large university at the present day. The teaching of the calculus and everything beyond was done by Benjamin Peirce and his son, the father at this period giving only the more advanced courses for the few upper classmen who elected them. He expected and received close and rapid attention in class, and hard, though not extensive, work outside. We read his *Analytic Mechanics*, Briot and Bouquet on *Elliptic Functions*, Tait and Hamilton on *Quaternions;* while his direct instruction consisted mainly, but not wholly, in solving problems by writing on the blackboard that covered the end of the room a series of equations which we copied into our notebooks.

As soon as he had finished the problem or filled the blackboard he would rub everything out and begin again. He was impatient of detail, and sometimes the result would not come out right; but instead of going over his work to find the error, he would rub it out, saying that he had made a mistake in a sign somewhere, and that we should find it when we went over our notes.

Described in this way it may seem strange that such a method of teaching should be inspiring; yet to us it was so in the highest degree. We were carried along by the rush of his thought, by the ease and grasp of his intellectual movement. The inspiration came, I think, partly from his treating us as highly competent pupils, capable of following his line of thought even through errors in transformations; partly from his rapid and graceful methods of proof, which reached a result with the least number of steps in the process, attaining thereby an artistic or literary character; and partly from the quality of his mind which tended to regard any mathematical theorem as a particular case of some more comprehensive one, so that we were led onward to constantly enlarging truths. To those of us who have not pursued the study of mathematics since college days the substance of what he taught us has faded away, but the methods of thought, the attitude of mind and the mode of approach have remained precious possessions.

III. REMINISCENCES.

By Professor Emeritus W. E. BYERLY, Harvard University.

When I entered Harvard in 1867, a particularly unsophisticated freshman from New Jersey knowing absolutely no one in the college, Cambridge was a small straggling town. The inhabitants still spoke of visiting Harvard Square as going down to the village.

The Square itself was occupied by the hay scales and the town pump. The portion of the college yard east of University Hall was a hayfield, from which the University drew a modest annual profit. The dormitories, Massachusetts, Hollis, Stoughton, and Holworthy, were grouped in the neighborhood of the college pump, the water supply of all the students in the yard. Steam heating and plumbing were unknown.

The College was small; the Faculty was small, but distinguished and picturesque. "There were giants in those days," bearded giants mainly, though Agassiz and Child were beardless, Sophocles, Longfellow, Lowell, Asa Gray, Benjamin Peirce. There are giants in the faculty now, but they are more or less lost in the crowd. Then, poets, discoverers, philosophers, and seers, in soft hats and long cloaks, looked their parts, and we newly-fledged freshmen gazed at them with admiration and awe.

The appearance of Professor Benjamin Peirce, whose long gray hair, straggling grizzled beard and unusually bright eyes sparkling under a soft felt hat, as he walked briskly but rather ungracefully across the college yard, fitted very well with the opinion current among us that we were looking upon a real live genius, who had a touch of the prophet in his make-up.

When I knew him later in the class-room, I will not say as a teacher, for he inspired rather than taught, and one's lecture notes on his courses were apt to be chaotic, I always had the feeling that his attitude toward his loved science was that of a devoted worshipper, rather than of a clear expounder. Although we could rarely follow him, we certainly sat up and took notice.

I can see him now at the blackboard, chalk in one hand and rubber in the other, writing rapidly and erasing recklessly, pausing every few minutes to face the class and comment earnestly, perhaps on the results of an elaborate calculation, perhaps on the greatness of the Creator, perhaps on the beauty and grandeur of Mathematics, always with a capital M. To him mathematics was not the handmaid of philosophy. It was not a humanly devised instrument of investigation, it was Philosophy itself, the divine revealer of TRUTH.

I remember his turning to us in the middle of a lecture on celestial mechanics and saying very impressively, "Gentlemen, as we study the universe we see everywhere the most tremendous manifestations of force. In our own experience we know of but one source of force, namely will. How then can we help regarding the forces we see in nature as due to the will of some omnipresent, omnipotent being? Gentlemen, there must be a GOD."

At another time he was lecturing on his favorite subject, which was then beginning to attract the attention of mathematicians and philosophers, Hamilton's new calculus of quaternions, which he believed was going to be developed into a most powerful instrument of research. He must have been working recently on his "Linear Algebras" for he said that "of possible quadruple algebras the one that had seemed to him by far the most beautiful and remarkable was practically identical with quaternions,[1] and that he thought it most interesting that a calculus which so strongly appealed to the human mind by its intrinsic beauty and symmetry should prove to be especially adapted to the study of natural phenomena. The mind of man and that of Nature's God must work in the same channels."

In one of his lectures on the theory of functions he established the relation connecting π, e, and i, $e^{\pi/2} = \sqrt[i]{i}$, which evidently had a strong hold on his imagination. He dropped his chalk and rubber, put his hands in his pockets, and after contemplating the formula a few minutes turned to his class and said very slowly and impressively, "Gentlemen, that is surely true, it is absolutely paradoxical, we can't understand it, and we don't know what it means, but we have proved it, and therefore we know it must be the truth."

I have hinted that his lectures were not easy to follow. They were never carefully prepared. The work with which he rapidly covered the blackboard was very illegible, marred with frequent erasures, and not infrequent mistakes (he worked too fast for accuracy). He was always ready to digress from the straight path and explore some sidetrack that had suddenly attracted his attention, but which was likely to have led nowhere when the college bell announced the close of the hour and we filed out, leaving him abstractedly staring at his work, still with chalk and eraser in his hands, entirely oblivious of his departing class.

Outside of the class-room I used to see him at meetings of a little informal mathematical club, attended by the more advanced students, where he frequently took part in the discussions and was always alert and suggestive; and at meetings of the American Academy where he frequently took an active part in the informal debate on the paper of the evening, usually to the enlightenment or the discomfiture of the author.

The first meeting of the Academy I ever attended gave him an opportunity to show his remarkable ability to think clearly and quickly. The paper of the evening was a very elaborate one, describing the lecturer's investigations into the tides of The Gulf of Maine. An important member of the Coast Survey, he had been engaged all summer in hydrographic work at the mouth of the Bay of Fundy, but he confessed himself completely staggered by the phenomena he had

---

[1] Compare pages 15–16, 28 of this monograph.—R. C. A.

observed and had just described to us, which seemed to him absolutely inexplicable. At the close of the address Professor Peirce rose from his seat and began to ask leading questions. The lecturer, rather puzzled at first, began to answer them hesitatingly but soon discovered that step by step he was being led up to a theory that met all his difficulties and dissolved all his paradoxes. It was as pretty a piece of work as ever I saw done, and was manifestly entirely unrehearsed.

Benjamin Peirce, mathematician and mystic, was not always on the heights. Calling at his house one day to consult him on some abstruse problem I found him on all fours in the parlor playing bear with one of his grandchildren, and I was invited to take part in the game.[1]

In his personal relations with his students he was always courteous, kind, and helpful, if rather prone to overrate their ability and promise, and they reverenced and loved him.

IV. REMINISCENCES.

By Chancellor ARNOLD B. CHACE, Brown University.

I have very pleasant memories of Professor Benjamin Peirce. In the later seventies being desirous of taking up the study of quaternions, which were then beginning to be talked about, and having worked at them myself for a while, I decided that I needed some help—and, going one day to Cambridge, after making some inquiries, I called on Professor Peirce, introduced myself, and asked him if he would assist me. He received me very pleasantly and seemed much interested in my request. He had at that time retired from active service in the college, but, as he told me with reference to a recent request from the head of Radcliffe, which was just starting, he was glad to assist anyone who deserved it.

I went to his house one afternoon a week for nearly a year and, sitting in his pleasant study before an open fire, I would show him the work that I had done in the previous week, and he, an old man, and I, a young man, discussed quaternions and many other matters in a most friendly way.

He was one of the most stimulating men that I have ever known. I can picture him now with his large noble brow, his beautiful white hair, his flashing eyes, his animated but kindly face, and most inspiring personality.

[1] A quotation from Henry Cabot Lodge's *Early Memories* (New York, 1913, pp. 55–56) may be recalled in this connection: "Altogether he had a fascination which even a child felt, and all the more because he was full of humor, with an abounding love of nonsense, one of the best of human possessions in this vale of tears. I know that I was always delighted to see him, because he was so gentle, so kind, so full of jokes with me, and so 'funny.' As time went on I came as a man to know him well and to value him more justly, but the love of the child, and the sense of fascination which the child felt, only grew with the years."

Another germane quotation may be made from E. W. Emerson's *The Early Years of the Saturday Club* (Boston and New York, 1918, p. 102): "A pleasant reminiscence of the family life is given by his daughter, another instance of Leasts and Mosts in this remarkable man. Before breakfast he always went to walk with his younger children, now a delightful memory to them. This man who could divine and see remotest suns in space, amused his little ones by allowing no pin to hide from his eyes in the dust of the sidewalk;—'although he never seemed to be looking for them, he would suddenly stoop to pick up a pin. He had various 'pincushions'; one was the trunk of an elm tree near our gate, others on Harvard and Brattle streets. Those on Quincy and Kirkland streets are still standing.' "—R. C. A.

He was very enthusiastic in his belief that Hamilton's *Lectures on Quaternions* and *Elements of Quaternions* marked a most important step in the progress of mathematical science, a belief which I think has been fully justified in all our modern vector analysis. A recent writer has placed [1] these books of Hamilton's among the ten most important works in the development of mathematics.

It was at just about this date, in the winter 1878–79, that Professor Peirce delivered a course of lectures at the Lowell Institute entitled "Ideality of the Physical Sciences." In the sixth and last of these lectures he made the statement that "Ideality is preëminently the foundation of mathematics." It was in this lecture also, after stating that two geometers had computed independently of each other the elements of the orbit of a planet which would reconcile the apparent discrepancies in the orbit of the recently discovered planet Uranus, and had determined that this new planet was situated at a certain point in the heavens, he told how on December 23, 1846, Dr. Galle of Berlin directed his telescope at the designated spot and discovered the new planet Neptune. Professor Peirce then made the surprising statement that the planet so discovered by Dr. Galle was not in the place that had been figured out, but that there were two possible positions of the planet and that by a remarkable coincidence on the given night the two positions were in a straight line from the earth. I remember very well as I sat in his study that he repeated this story to me with much animation, and when I questioned him further about it he said he was sure of his reasoning, but the calculations were so long and laborious that he had never had the courage to go through them a second time.[2]

V. Biographical Sketch.[3]

By R. C. Archibald, Brown University.

Mathematical research in American Universities began with Benjamin Peirce. His influence on students and contemporaries was extraordinary; this is borne out by the "Reminiscences" given above. In September, 1924, President Lowell wrote also: "I have never admired the intellect of any man as much as that of Benjamin Peirce. I took every course that he gave when I was in College, and whatever I have been able to do intellectually has been due to his teaching more than to anything else."

Hence no apology is required for taking the greater part of an issue of the Monthly to exhibit the life and work of such a man, if information of this kind is not already easily available. An appreciative sketch appeared in this Monthly nearly thirty years ago,[4] but it seemed evident that something of a more com-

---

[1] In this Monthly, *1923*, 320.—R. C. A.

[2] Compare page 14 of this monograph.—R. C. A.

[3] The gist of this section and of the next was given in a paper on Benjamin Peirce read at a joint session of the History of Science Society, Section L of the American Association for the Advancement of Science, and of the Mathematical Association of America, Washington, D. C., January 1, 1925.

[4] "Benjamin Peirce" by F. B. Matz, in this Monthly, *1895*, 173–179; also in *A Mathematical Solution Book* by B. F. Finkel, fourth ed., Springfield, Mo., 1902, pp. 524–528.

BENJAMIN PEIRCE, 1853 (?)

*From a painting by J. A. Ames in
the possession of Harvard University.*

prehensive nature should be attempted, and that sources of further information should be indicated for the present generation and for the future historian of American mathematics. Previously [1] there has been no adequate indication of the extent of Peirce's publications; even the lists of his periodical articles in the *Royal Society Catalogue of Scientific Papers*, and in "Poggendorff," [2] are by no means complete. Furthermore, critical estimates of Peirce's notable work in linear associative algebra, in connection with the problem of the discovery of Neptune, and in other fields, are not readily to be found by the average inquirer.

Benjamin Peirce was born at Salem, Mass., April 4, 1809, and died at Cambridge, Mass., October 6, 1880. He was descended from John Pers, a weaver of Norwich, Norfolk Co., England, who emigrated to this country in 1637. His father was Benjamin Peirce (1778–1831), who graduated from Harvard University in 1801, represented Salem in the lower branch of the legislature for several years and was later sent to the state senate. For the last five years of his life he was librarian at Harvard University; his history of the University was published after his death.

Benjamin Peirce entered Harvard University in 1825 and graduated in the class of 1829, with membership in the Phi Beta Kappa Society. Oliver Wendell Holmes, James Freeman Clarke, educator and prolific author, and George T. Bigelow and Benjamin R. Curtis, eminent jurists, were classmates. For the two years immediately after graduation, Peirce was associated with George Bancroft as teacher at the famous Round Hill School, Northampton, Mass. In 1831 he was appointed tutor in mathematics at Harvard College and

---

[1] Some of the best printed sources of information concerning Peirce's ancestry, life and work are: (1) F. C. Peirce, *Peirce Genealogy*, Worcester, 1880; (2) R. S. Rantoul, *Historical Colls. Essex Institute*, vol. 18, 1881, pp. 161–176; (3) *Proc. Amer. Acad. Arts and Sciences*, vol. 16, 1881, pp. 443–454, by [H. A. Newton]. This appeared in slightly different form in *Amer. Jl. Sci.*, vol. 122, 1881, pp. 167–178; (4) *Benjamin Peirce . . . A Memorial Collection*, edited by M. King, Cambridge, Mass., 1881, 64 pp. [includes sketch by Thomas Hill from *The Harvard Register*, May, 1880; editorials and sketches from various newspapers and periodicals; sermons by A. P. Peabody, T. Hill and C. A. Bartol; and poems by Oliver Wendell Holmes, G. Thwing and T. W. Parsons. The poem by Holmes appeared originally in the *Atlantic Monthly*, vol. 46, 1880, p. 823; also in *Kansas City Review*, vol. 4, p. 510]; (5) *Mo. Notices Royal Astr. Soc.*, vol. 41, 1881, pp. 191–193; (6) *Proc. Royal Soc. Edinb.*, vol. 22, 1882, pp. 739–743, by Simon Newcomb; (7) "The Services of Benjamin Peirce to American mathematics and astronomy" by J. J. See, *Popular Astronomy*, vol. 3, 1895, pp. 49–57; (8) *Encyclopædia Britannica*, eleventh edition, vol. 11, 1911; (9) F. Cajori, *A History of Mathematics*, second ed., 1919, pp. 338–339, etc.; and (10) E. W. Emerson, *The Early Years of the Saturday Club, 1855–1870*, Boston and New York, 1918, pp. 96–109. Other references will be given in the following pages. Sketches of minor importance occur in: (1) F. S. Drake, *Dictionary of American Biography*, Boston, 1872; (2) *The Harvard Book, a series of historical and biographical and descriptive sketches* by various authors collected and publ. by F. O. Vaille and H. A. Clark, vol. 1, Cambridge, 1875, pp. 104, 172–173; (3) Appleton's *Cyclopædia of American Biography*, New York, vol. 3, 1888; and (4) *National Cyclopædia of American Biography*, New York, vols. 8, 9, 10, 1898, 1907, 1909.

A considerable quantity of Benjamin Peirce's manuscripts and correspondence was presented to the American Academy of Arts and Sciences in 1913. This collection is soon to be augmented by many other letters of great value which have been in the possession of Peirce's grandson, Benjamin P. Ellis of Cambridge, Mass. I am much indebted to Mr. Ellis for allowing me free access to this material.

[2] J. C. Poggendorff, *Biographisch-Literarisches Handwörterbuch zur Geschichte der exacten Wissenschaften*, Leipzig, vols. 2, 3, 1863, 1898.

was in full charge of the mathematical work. In 1833 he received the A.M. degree from Harvard, was appointed professor of mathematics and natural philosophy, and was married to Sarah H. Mills of Northampton. His academic title was changed to that of professor of astronomy and mathematics in 1842.

Professor Peirce had four sons and a daughter. One son, Benjamin Mills (1844–70), was a mining engineer; another, Herbert Henry Davis [1] (1849–1916), was a diplomat.[2] The eldest, James Mills (1834–1906), was assistant professor of mathematics at Harvard 1861–69, and professor from 1869 till his death; he was also dean of the graduate school of Arts and Sciences 1890–95, and of the faculty of Arts and Sciences 1895–98. But the son who seemed largely to inherit his father's intellectual powers [3] was Charles Santiago Saunders (1839–1914) who was lecturer at Harvard, in philosophy and logic, 1869–71. Of these brothers, the brilliant Benjamin Osgood Peirce (1854–1914) was a second cousin once removed.[4]

Professor Benjamin Peirce was honored in various ways both in this and in other countries. He was elected a Member of the American Philosophical Society in 1842; an Associate of the Royal Astronomical Society, London, 1850; a Foreign Member (limited to 50) of the Royal Society of London in 1852; an Honorary Member of the State Historical Society of Wisconsin, 1854; a Fellow of the American Academy of Arts and Sciences in 1858; an Honorary Fellow of the University of St. Vladimir, at Kiev, Russia (now Ukrainia), 1860; a Corresponding Member of the British Association for the Advancement of Sciences in 1861; an Honorary Fellow (limited to 36) of the Royal Society of Edinburgh in 1867; and a Correspondent in the mathematical class of the Royal Society of Sciences at Göttingen in 1867. He received the degree of LL.D. from the University of North Carolina in 1847, and from Harvard [5] in 1867. He was: One of a committee of five appointed by the American Academy of Arts and Sciences to draw up a "Program for the Organization of the Smithsonian Institution," 1847; Consulting Astronomer for the Nautical Almanac, 1849–67; Director of the

---

[1] Probably named after Admiral Charles Henry Davis (1807–77), who married Benjamin Peirce's wife's sister.

[2] See *National Cyclopædia of American Biography*, vol. 10, p. 449, and vol. 9, p. 539.

[3] Papers of C. S. S. Peirce are referred to in such works as: B. Russell, *The Principles of Mathematics*, vol. 1, Cambridge, 1903; C. I. Lewis, *A Survey of Symbolic Logic*, Berkeley, 1918; and F. Enriques, *Per la Storia della Logica*, Bologna, 1922. See, also, E. W. Davis, "Charles Peirce at Johns Hopkins," *The Mid-West Quarterly*, New York, vol. 2, pp. 48–56; it is here stated that Sylvester considered C. S. Peirce "a far greater mathematician than his father."

[4] At Harvard he was instructor in mathematics, 1881–84, then assistant professor of mathematics and physics, 1884–88, and finally professor of mathematics and natural philosophy, 1888–1914.

[5] In a letter dated July 29, 1867, the president of Harvard College, who was also somewhat of a mathematician, wrote as follows:

"I have the honor of informing you that the University, on Commencement Day, conferred upon you the degree of Doctor of Laws in recognition of the transcendent ability with which you have pursued mathematical physical investigations, and in particular for the luster which she has herself for so many years borrowed from your genius.

"With the sincerest regard,

"Very truly and gratefully yours,

"THOMAS HILL."

longitude determinations of the United States Coast Survey, 1852–67; Member of the Scientific Council (J. Henry, A. D. Bache, B. Peirce) of the Dudley Observatory, Albany, 1855–58; Superintendent of the Coast Survey, February 26, 1867, to February 16, 1874, while continuing to serve as professor at Harvard; consulting geometer of the Survey,[1] 1874–80; President of the American Association for the Advancement of Science, 1853, and elected Fellow, 1875; Chairman of the Department of Education of the American Social Science Association, 1869–72, acting president [2] in 1878, and vice-president in 1880; One of the fifty incorporators of the National Academy of Sciences, one of the nine members of the committee of organization, and chairman of the mathematics and physics class,[3] 1863; Director of the expedition to Sicily to observe the eclipse of the sun, December, 1870; Coöperating Editor of the *American Journal of Mathematics*, volume 1, 1878; Special Lecturer on physical philosophy at the Concord Summer School of Philosophy and Literature, 1879; Lecturer at the Lowell Institute, 1879, and at the Peabody Institute, 1880.

In a recently published article, Professor Coolidge pointed out [4] that before the time of Benjamin Peirce it never occurred to anyone that mathematical research "was one of the things for which a mathematical department existed. Today it is a commonplace in all the leading universities. Peirce stood alone— a mountain peak whose absolute height might be hard to measure, but which towered above all the surrounding country." In his publications [5] and papers before scientific bodies, Peirce touched on a wide range of topics.

Of his eleven works, in twelve volumes, six were elementary texts, some of which went through several editions. The first, on plane trigonometry, appeared in 1835, and the second and third, on spherical trigonometry and sound, in 1836. The seventh work, in two volumes (1841–46), dealt with analytic geometry,

[1] This post was held while retaining his professorship. He was appointed "consulting geometer" with compensation at the rate of $4,000 per annum, and subsistence at the same rate per diem as was allowed the late Hydrographic Inspector." For an account of Peirce's connection with the Survey, see *Centennial Celebration of the United States Coast and Geodetic Survey, April 5 and 6, 1916*, Washington, 1916, p. 137. T. C. Mendenhall here remarks, "As a genius in mathematics and astronomy he is easily a star of first magnitude in the Coast Survey galaxy." In an account of the centennial celebration in the *Scientific Monthly*, vol. 3, 1916, p. 616, the story is told that at a meeting of the National Academy of Sciences he spent an hour filling the blackboard with equations, and then remarked, "There is only one member of the Academy who can understand my work and he is in South America." Was this B. A. Gould? Hilgard was the managing head of the Survey during Peirce's administration; see anonymous note by S. Newcomb, *Nation*, March 5, 1874, vol. 13, p. 157.

[2] He declined to take the titular office of president offered to him in this year although performing all the duties of the office. For a sketch of Professor Peirce including an account, by F. B. Sanborn, of his relations to the American Social Science Association, see *Journal of Social Science*, no. 12, 1880, pp. ix–xi. For the title of his address delivered before the Association in 1878, see the List of Peirce's Writings in the next Section.

[3] Compare *A History of the First Half-Century of the National Academy of Sciences, 1863–1913*, Washington, 1913, pp. 9, 10, 20, 21, 23, 27, 168–171, 215, 223, and 256. Peirce was one of the first sixteen to read papers before the Academy, January, 1864.

[4] "The Story of Mathematics at Harvard" by J. L. Coolidge, *Harvard Alumni Bulletin*, January 3, 1924, vol. 26, p. 376.

[5] A list of these, which probably closely approximates to completeness, is given in Section VI of this monograph.

differential and integral caculus, and differential equations. His other volumes were: *Tables of the Moon* (1853–56); a notable work on *Analytic Mechanics* (1855), the remarkably original *Linear Associative Algebra* (1870), and the posthumous volume of lectures, *Ideality in the Physical Sciences* (1881).

About one quarter of the titles of Peirce's publications relate to topics of pure mathematics and three quarters to questions mainly in the fields of astronomy, geodesy and mechanics. His first publications, when only sixteen years of age, were solutions of problems in algebra and mechanics. Very early in life, possibly through having Ingersoll Bowditch as schoolmate, Peirce had the good fortune to become acquainted with Ingersoll's father, Dr. Nathaniel Bowditch [1] (1773–1838), author of *The New American Practical Navigator* (which has gone through so many editions in the last hundred years), and translator of Laplace's celestial mechanics. During the ten years before he was thirty, Peirce revised and corrected the proof sheets of this translation. Among other works, Peirce contributed an original notable result regarding perfect numbers; gave certain methods of determining the number of real roots of equations applicable to transcendental as well as to algebraic equations; made an important advance in the treatment of Kirkman's school-girl problem; [2] discussed a new binary system of arithmetic; wrote on probabilities at the three-ball game of billiards, on the extension of Lagrange's theorem for development of functions, and on transformation of curves; and supplemented his volume on associative algebra by a memoir on the uses and transformations of linear algebras. Apart from volumes already referred to, his publications on applied mathematics included: various papers on the perturbations of Neptune and Uranus; a mathematical treatment (also translated into German) of the problem of Saturn's rings, leading to the result that the rings were fluid; [3] a note upon the conical pendulum; papers on the relation between the elastic curve and the motion of the pendulum, on a criterion for the rejection of doubtful observations, on the catenary on a vertical right cone, on the internal constitution of the earth, and on a mathematical investigation of the fractions which occur in phyllotaxis.

The concluding characteristic paragraph of this last paper is as follows:

"May I close with the remark, that the object of geometry in all its measuring and computing, is to ascertain with exactness the plan of the great Geometer, to penetrate the veil of material forms, and disclose the thoughts which lie beneath them? When our researches are successful, and when a generous and heaven-eyed inspiration has elevated us above humanity, and raised us triumphantly into the very presence, as it were, of the divine intellect, how instantly and entirely are human pride and vanity repressed, and, by a single glance at the glories of the infinite mind, are we humbled to the dust." [4]

---

[1] The dedication of Peirce's *Analytic Mechanics* is as follows: "To the cherished and revered memory of my master in science, Nathaniel Bowditch, the father of American geometry, this volume is inscribed."

[2] Sylvester referred to this treatment as "the latest and probably the best" (*Philosophical Magazine*, 1861, vol. 21, p. 520; also *Collected Mathematical Papers of . . . Sylvester*, vol. 2, 1908, p. 276). See also my notes on this title in the next Section.

[3] Compare S. Newcomb, *Popular Astronomy*, 1879, p. 358; also Newcomb-Engelmann, *Populäre Astronomie*, 5th ed., 1914, p. 429.

[4] Another quotation may be given to illustrate Peirce's manner of thought to which reference has been made above in the "Reminiscences." The following are concluding sentences from a

While Peirce read before scientific societies many papers concerning his investigations, the printed reports of them are unfortunately often mere abstracts. "His mind moved with great rapidity, and it was with difficulty that he brought himself to write out even the briefest record of its excursions."

"His elementary books were remarkable for their condensation. In the geometry, especially, the short and terse and comprehensive forms of mathematical thought and expression, natural to the mathematician, were substituted for the minute demonstrations of Euclid. Free use was also made of infinitesimals." [1]

In order to bring out more clearly the place Peirce occupies in the development of American mathematics it seems desirable to comment further on four of the subjects which he discussed in a notable manner: 1. criterion for the rejection of doubtful observations; 2. perturbations of Uranus and the discovery of Neptune; 3. analytic mechanics; 4. linear associate algebra.

1. "Peirce's criterion," as the term has appeared in scientific literature, had as its object the solution of a delicate and practically important problem of probability; this problem is: "Being given certain observations of which the greater part is to be regarded as normal and subject to the ordinary law of error adopted in the method of least squares, while a smaller unknown portion is abnormal, and subject to some obscure source of error, to ascertain the most probable hypothesis as to the partition of the observations into normal and abnormal." This criterion has been regarded as one of Peirce's best contributions to science. In volume 46 (St. Petersburg, 1898) of the great Russian encyclopædia (based on "Brockhaus") it is especially referred to in a ten-line biographical notice of Peirce.

The excessive brevity of Peirce's statement concerning the criterion, when it appeared in 1852, resulted in frequent misunderstandings. The tables which Gould published three years later [2] facilitated its application. But it was not till 1878 that Peirce somewhat remedied his original statement by giving fuller explanations. In 1920, however, R. M. Stewart proved the statement fallacious.[3]

The criterion and its application are set forth at length in W. Chauvenet's *Manual of Spherical and Practical Astronomy* [4] (1868), and a paragraph is devoted to it in W. S. Jevons's *Principles of Science* [5] (1877). An illustration of a recent work where application of the criterion is suggested is H. M. Wilson's *Topographic, Trigonometric and Geodetic Surveying* [6] (1912).

paper on Saturn's rings: "But in approaching the forbidden limits of human knowledge, it is becoming to tread with caution and circumspection. Man's speculations should be subdued from all rashness and extravagance in the immediate presence of the Creator. And a wise philosophy will beware lest it strengthen the arms of atheism, by venturing too boldly into so remote and obscure a field of speculation as that of the mode of creation which was adopted by the Divine Geometer."

[1] S. Newcomb, *Royal Soc. Edinb., Proc.*, vol. 22, p. 739.

[2] B. A. Gould, "On Peirce's criterion for the rejection of doubtful observations with tables for facilitating its application," *Astron. Jl.*, vol. 4, pp. 81–87. G. B. Airy expressed himself as believing Peirce's criterion defective in its foundation and illusory in its results (*Astron. Jl.*, vol. 4, pp. 137–138, 1856); Joseph Winlock showed (pp. 145–146) that his argument was wholly unsound. Compare articles by Stone and Glaisher in *Mo. Notices R. Astr. Soc.*, vols. 28, 33–35.

[3] *Popular Astronomy*, vol. 28, pp. 2–3. See also J. L. Coolidge, *An Introduction to Mathematical Probability*, Oxford, 1925, pp. 126–127.

[4] Volume 2, fourth edition, 1868, pp. 558–566, 596–599.

[5] London, second edition, p. 391.   [6] New York, third edition, pp. 604–606.

2. The computation of the general perturbations of Uranus and Neptune was the first work to extend Peirce's reputation. Simon Newcomb's compact statement [1] in this connection is here reproduced with two added footnotes:

"The formulæ to which he was led were published in the first volume of the *Proceedings of the American Academy*,[2] but were accompanied by no description of his process. Subsequent investigations, however, showed them to have been remarkably accurate. In his views of the discrepancy between the mean distance of Neptune as predicted by Leverrier, and as deduced from observations, he was less fortunate, although when due consideration is given to Leverrier's conclusions, there was much plausibility in the position taken by Peirce. As the subject has frequently been discussed without a due comprehension of all the circumstances, a brief review of them may be appropriate.[3]

"Leverrier, from his researches, found for the mean distance of the disturbing planet, 36.1539, and a consequent period of 217 years. He also announced that the limits of the mean distance which would satisfy the observed perturbation of Uranus were 35.04 and 37.90. He founded this conclusion on a supposed inadmissible increase of the outstanding differences between theory and observation, as the mean distance was diminished below 35. But when the planet was discovered, its mean distance was found to be only 30; and yet the observations of Uranus were as well satisfied as by Leverrier's hypothetical planet. It was, therefore, an expression of Peirce's high confidence in the accuracy of Leverrier's conclusions that led him to announce that there were two solutions to the problem; the one being that found by Leverrier, and the other that corresponding to the actual case. He also sought to show a cause for the two solutions in a supposed discontinuity in the form of the perturbations, when the period was brought to the point at which five revolutions of Uranus would be equal to two of Neptune. As a matter of fact, however, it has been shown by Professor Adams that there is no such discontinuity in the actual perturbations during the limited period; from which it would follow that Leverrier must have made a mistake in tracing out the conclusions which would follow when the mean distance of the disturbing planet was diminished."

H. H. Turner has also published a very readable account of this matter in his *Astronomical Discovery* (London, 1904).

3. As to Peirce's *Analytic Mechanics*, Simon Newcomb referred to it [4] as "the most characteristic as well as the most extensive of his works." Then he continues: "The exposition of dynamical concepts in the first forty pages is pleasant reading for one already acquainted with the subject, but that a student beginning the subject could understand it without clearer distinction of definitions, axioms, and theorems seems hardly possible." In his later years Peirce often said he wanted to rewrite his *Mechanics* and introduce quaternions into it. Sir

---

[1] *Roy. Soc. Edinb., Proc.*, vol. 22, pp. 740–741.

[2] Published in 1848. It is related that when, in 1846, Peirce announced in the American Academy that Galle's discovery of Neptune in the place predicted by Leverrier was a happy accident, the President, Edward Everett, "hoped the announcement would not be made public: nothing could be more improbable than such a coincidence."—"Yes," replied Peirce, "but it would be still more strange if there was an error in my calculations,"—a confident assertion which the lapse of time has vindicated. In this connection, it is noteworthy that Peirce was not made a fellow of the American Academy till many years after he had been honored by the American Philosophical Society and two foreign bodies. In 1878 Peirce sent in his resignation as a Fellow of the Academy, but this was never accepted.

[3] A full statement of Peirce's views, which he maintained to the last, is given by J. M. Peirce, on pages 200–211 of B. Peirce's *Ideality in the Physical Sciences*. It was 28 years after Peirce's criticisms of the work of Leverrier and Adams were published (1848), that is, 1876, that J. C. Adams made a reply, *Journal de Mathématiques* (Liouville), vol. 41, 1876; see also *The Scientific Papers of John Couch Adams*, vol. 1, 1896, pp. xxxiii, 57, 64.

[4] *Royal Society Edinb., Proc.*, vol. 22, p. 742.

Thomas Muir has given [1] an analysis of the section on determinants and functional determinants. The analysis is as follows:

"At the outset of his Tenth Chapter, which deals with the integration of the differential equations of motion, Peirce feels the need for making his reader acquainted with the properties of functional determinants. He accordingly gives as a preparation a brief account (§§ 327–348, pp. 172–183) of determinants in general, and then expounds within the space of sixteen broad-margined pages the main theorems of Jacobi's 'De determinantibus functionalibus.' The treatment of the original is free and masterly, the order being altered with good effect. For example, Jacobi's incorrectly stated proposition is brought forward to occupy the second place, the enunciation being *If either* (i.e., *any one*) *of the given functions contains any of the other functions, these* (latter) *functions may be regarded as constant in finding the functional determinant.* There is thence deduced Jacobi's last proposition of all, namely, that expressing the determinant as a single product: and this in turn is used to discuss the connection between the vanishing of the determinant and interdependence of the functions.

"Had Peirce's exposition been less condensed and been published as part of an ordinary textbook of determinants, its value at this time to English-speaking students would have been considerable."

4. There seems to be no question that his *Linear Associative Algebra* [2] was the most original and able mathematical contribution which Peirce made. He himself held the work in high esteem; on April 4, 1870, he wrote in the introduction, "This work has been the pleasantest mathematical effort of my life. In no other have I seemed to myself to have received so full a reward for my mental labor in the novelty and breadth of the results." In his *Synopsis of Linear Associative Algebra* (published by the Carnegie Institution in 1907) J. B. Shaw characterized the work as "really epoch-making," and devoted a number of pages (52–55, 101–106) to formulating the main results. While the monograph attracted wide and favorable comment in England and America,[3] continental investigators on the subject (1889–1902) did not give Peirce the credit which his results and methods deserved. Adverse criticism had been "due in part to a misunderstanding of Peirce's definitions, in part to the fact that certain of Peirce's principles of classification are entirely arbitrary and quite distinct in statement from those used by Study and Scheffers,[4] in part to Peirce's vague and in some cases unsatisfactory proofs, and finally to the extreme generality of the point of view from which his memoir sprang, namely a 'philosophic study of the laws of algebraic operation.'" In order that Peirce's work should receive due recognition, H. E. Hawkes discussed and answered the following questions: [5]

[1] T. Muir, *The Theory of Determinants*, London, vol. 2, 1911, p. 251.

[2] The first sentence of the work is often quoted. It is: "Mathematics is the science which draws necessary conclusions." On page 5 we find a reference to the "mysterious formula" $i^{-i} = e^{\pi/2} = 4.810477381$. Compare this MONTHLY, *1921*, 115–121.

[3] The substance of the work was reviewed by William Spottiswoode in his retiring presidential address delivered before the London Mathematical Society in 1872 ("Remarks on Some Recent Generalizations in Algebra," *London, Math. Soc., Proc.*, vol. 4, 1873, pp. 147–164). Peirce refers to this (1875) as a "fine, generous, and complete analysis." See also Cayley, *Collected Mathematical Papers*, vol. 11, pp. 457–8; vol. 12, pp. 60–71, 106, 303, 459, 465. The first reference is to Cayley's address before the British Association in 1883, when he spoke of "the valuable memoir by the late Benjamin Peirce." At the last reference Cayley writes (1887): "the general theory of associative linear forms is treated in a very satisfactory manner in Peirce's memoir."

[4] A complete list of references may be found in Shaw's work. So also for references to other works developed from Peirce's ideas.

[5] "Estimate of Peirce's Linear Associative Algebra," *Amer. Jl. Math.*, vol. 24, 1902, pp. 87–95; "On Hypercomplex Number Systems," *Amer. Math. Soc., Trans.*, vol. 3, 1902, pp. 312–330.

(1) What problem did Peirce attack, and to what extent did he solve it? (2) What relation does this problem bear to that treated by Study and Scheffers? (3) To what extent do Peirce's methods assist in the solution of that problem? In part Hawkes summed up his conclusions as follows:

"We can now state precisely the problem that Peirce set for himself. He aimed to develop so much of the theory of hyper-complex numbers as would enable him to enumerate all inequivalent, pure, non-reciprocal number systems in less than seven units. The relation to the problem treated by Scheffers is plain if we remember that the first two of Peirce's principles of classification are identical with those of Scheffers, and the other three are only slightly modified. Peirce solved this problem completely. The theorems stated by him are in every case true, though in some cases his proofs are invalid."

Hawkes showed also that by using Peirce's principles as a foundation, we can deduce a "method more powerful than those hitherto given," by such writers as Study and Scheffers, for enumerating all number systems of the type considered by Scheffers. Since Study is the author of the article on "Theorie der gemeinen und höheren complexen Grössen" in the *Encyklopädie der math. Wissenschaften*, one is not surprised to find his references to Peirce so wholly inadequate.[1]

Benjamin Peirce died in 1880 in the seventy-second year of his life and in the fiftieth of his service in the University. The teaching of mathematics at Harvard during this half century has been described by Florian Cajori[2] and J. K. Whittemore.[3] Other information of interest, supplementing the "Reminiscences" given above, may be found in the printed Reports of the presidents of the University during those years, in the 1829 Class Records (Harvard Univ. Library), and in the first volume of the *Annals of the Harvard Observatory*.

In the account of "How I was Educated," Edward Everett Hale, '39, wrote:[4] "The classical men made us hate Latin and Greek; but the mathematical men (such men! Pierce [*sic*] and Lovering) made us love mathematics, and we shall always be grateful to them." Lovering taught mathematics and natural philosophy at Harvard, 1838–88. In another place Hale wrote:[5] "I had but four teachers in college,—Channing, Longfellow, Peirce and Bachi. The rest heard me recite but taught me nothing."

Colonel Henry Lee has written of Peirce as follows in *The Harvard Book*:[6]

"Why we should have given him the affectionate diminutive name of 'Benny' I cannot say, unless as a mark of endearment because he could fling the iron bar upon the Delta farther than any undergraduate, or perhaps because he always thought the bonfire or disturbance outside the college

[1] Volume 1, pp. 159 and 167. In Cartan's form of the article, *Encyclopédie des Sciences Mathématiques*, tome 1, vol. 1, a very different presentation is found; see, for example, pp. 369, 401–2, 417, 422–5. While scores of trivialities are reviewed in *Jahrbuch über die Fortschritte der Mathematik* for 1881, absolutely nothing is given concerning Peirce's notable work.

[2] F. Cajori, *The Teaching and History of Mathematics in the United States*, Washington, 1890, pp. 133–148, 278, 397. See also J. L. Coolidge, Story of math. at Harvard, *l. c.*, pp. 372–376.

[3] In a sympathetic sketch of "James Mills Peirce," *Science*, n.s., vol. 24, 1906, pp. 40–48. Rather curiously J. M. Peirce, who died in 1906, was, as his father also, in the seventy-second year of his life and in the fiftieth of his service in the University.

[4] *Forum*, vol. 1, 1886, p. 61.

[5] *Outlook*, June 4, 1889, vol. 59, p. 316; also in E. E. Hale, *James Russell Lowell and his Friends*, Boston, 1899, p. 128.

[6] Volume 1, Cambridge, 1875, p. 104.

grounds, and not inside, and conducted himself accordingly. His softly lisped *sufficient* brought the blunderer down from the blackboard with a consciousness of failure as overwhelming as the severest reprimand. There was a delightful abstraction about this absorbed mathematician which endeared him to the students, who hate and torment an instructor always on the watch for offences, and which confirmed the belief in his peculiar genius."

Peirce's force and judgment in a great emergency are shown in the following anecdote by one who was present:[1]

Jenny Lind's last concert of the original series, given under the auspices of Phineas T. Barnum, was given at the hall over the Fitchburg Railroad Station. Tickets were sold without limit,—many more than the hall could hold,—and there was every prospect of a riot. Barnum had taken the precaution to leave for New York. I got about one-third up the main aisle, but could get no farther. Just ahead of me was Professor Peirce. The alarm was increasing. The floor seemed to have no support underneath, but to hang over the railroad track by steel braces from the rafters above. Would it hold? The air was stifling and windows were broken, with much noisy crashing of glass, in order to get breath. Women were getting uneasy. And there was no possibility of escape from a mass of human beings so packed together. We knew, from the conductor's baton, that the orchestra was playing, but no musical sound reached us. Professor Peirce mounted a chair. Perfect silence ensued as soon as he made himself seen. He stated, very calmly, certain views at which he had arrived after a careful study of the situation. The trouble was at once allayed. Jenny Lind recovered her voice and the concert went on to its conclusion."

Another contemporary has made this record:[2]

"He was among the first to read any new and noteworthy poem [3] or tale, to hear a new opera or oratorio; and his judgment and criticism upon such matters was keen and original. His interest in religious themes was deep, but it was in the fundamental doctrines rather than in the debates of sectarians; he was a devout believer in Christianity, but held to no established creed."

Among Peirce's students who afterwards became eminent were Simon Newcomb and George W. Hill. In his *Reminiscences of an Astronomer*,[4] Newcomb has happily hit off some of Peirce's most striking characteristics as follows:

"Professor Peirce was much more than a mathematician. Like many men of the time, he was a warm lover and a cordial hater. It could not always be guessed which side of a disputed question he would take; but one might be fairly sure that he would be at one extreme or the other. As a speaker and lecturer he was very pleasing, neither impressive nor eloquent, and yet interesting from his earnestness and vivacity. For this reason it is said that he was once chosen to enforce the views of the university professors at a town meeting, where some subject of interest to them was coming up for discussion. Several of the professors attended the meeting, and Peirce made his speech. Then a townsman rose and took the opposite side, expressing the hope that the meeting would not allow itself to be dictated to by these nabobs of Harvard College. When he sat down, Peirce remained in placid silence, making no reply. When the meeting broke up, some one asked Peirce why he had not replied to the man. 'Why! did you not hear what he called us? He said we were nabobs! I so enjoyed sitting up there and seeing all the crowd look up to me as a nabob that I could not say one word against the fellow.'"

An estimate by one who was not a scientist may be added. In a centennial address Wendell Phillips referred [5] to Peirce as "the largest natural genius, the man of the deepest reach and firmest grasp and widest sympathy, that God has

---

[1] E. W. Emerson, *The Early Years of the Saturday Club*, Boston and New York, 1918, pp. 100–101.

[2] *Nation*, New York, October 14, 1880.

[3] Judge Addison Brown wrote that "Professor Peirce seemed a poet in a mathematical dream, his mind so preoccupied, as it were gazing at the stars" (*Annals of the Harvard Class of 1852*, by G. W. Edes, Cambridge, 1922, p. 322; there are references to Peirce on several other pages).

[4] Boston, 1903, pp. 77–78; see also pp. 276–277.

[5] *The Scholar in a Republic. Address at the Centennial Anniversary of the Phi Beta Kappa Society of Harvard College*, Boston, 1881, p. 13.

given to Harvard in our day,—whose presence made you the loftiest peak and farthest outpost of more than mere scientific thought,—the magnet who, with his twin Agassiz, made Harvard for forty years the intellectual Mecca of forty States."

Andrew P. Peabody, '26, who was preacher to the University and professor of Christian morals at Harvard for the last two decades of Peirce's service, has devoted several pages of his *Harvard Reminiscences* [1] to Peirce. In referring to the last few years of Peirce's life he tells how he

"had for his pupils only young men who were prepared for profounder study than ever entered into a required course, or a regularly planned curriculum; but he never before taught so efficiently, or with results so worthy of the mind and heart and soul, which he always put into his work. His students were inflamed by his fervor, and started by him on the eager pursuit of the eternal truth of God, of which mathematical signs and quantities are the symbols."

It was not alone "young men" whom Peirce was willing to direct, as the following extract from a letter [2] written in 1879 shows: "I will do the same for the young women that I do for the young men. I shall take pleasure in giving gratuitous instruction to any person whom I find competent to receive it. I give no elementary instruction, but only in the higher mathematics."

The pall-bearers at Peirce's funeral were President Eliot, Ex-President Thomas Hill, C. P. Patterson (Superintendent of the Coast Survey), J. J. Sylvester, [3] J. Ingersoll Bowditch, Simon Newcomb, Joseph Lovering, Andrew P. Peabody, and his classmates, James F. Clarke and Oliver Wendell Holmes.

At a meeting of the President and Fellows of Harvard College, October 11, 1880, the entry made upon the records regarding Peirce stated that

"The University must long lament the loss of an intelligence so rare, an experience so rich, and a personal influence so strong, as his.

"As a teacher, he inspired young minds with a love of truth, and touched them with his own enthusiasm; as a man of science, his attainments and achievements and his public services have reflected honor upon the University and the country."

What has been indicated above, coupled with the bibliography which follows, provides material necessary for forming an intelligent opinion as to the activities, personality, brilliant and powerful mind, and "wonderfully stimulating influence" of one of the most eminent and original scientists that America produced in the last century. In this country Peirce was the leading mathematician of his time, and a pioneer in achieving notable mathematical research, some of it anticipating work by well-known Europeans of later date. It is interesting to speculate as to the possible publication harvest if Peirce had been able throughout his career constantly to meet his mathematical equals or peers, and if he had

---

[1] Boston, 1888, pp. 180–186.

[2] From a letter in the possession of D. E. Smith of Columbia University.

[3] "When Professor Sylvester was called [1876] to the chair of mathematics in the Johns Hopkins University, Professor Peirce of Harvard, being asked what he thought would be the opinion of American mathematicians respecting the new appointment, replied that no American mathematician had a right to have any opinion on the subject, except himself, and one of his old pupils, a distinguished professor of mathematics in one of our leading colleges." L. A. Wait, "Advanced Instruction in American Colleges," *The Harvard Register*, vol. 3, p. 127, 1880; see also page 119.

had a capable disciple always at hand to put his ideas on paper in a form suitable for publication.

———————

In a passport dated May 14, 1860, Peirce is described as aged 51, of height 5 feet 7¾ inches, and with high forehead, hazel eyes, straight nose, regular mouth, round chin, brown hair, light complexion and oval face. A ticket dated Crystal Palace, London, October 5, 1873, gives his weight as 190 pounds.

Portraits of Peirce may be seen in the following sources:

Daguerreotype in possession of Peirce's grandson, Mr. Benjamin P. Ellis. This is the earliest known portrait of Peirce and is now reproduced for the first time. Probably it dates from about 1845.

Painting in the Faculty Room, University Hall, Harvard University, executed by J. A. Ames (1816–1872) about 1853(?). It was Mrs. Peirce's favorite portrait, and was presented to Harvard after her death in 1888. Reproduced herewith; also in *Harvard Alumni Bulletin*, January 3, 1924, vol. 26, p. 375.

Painting executed in 1857 by Daniel Huntington (1816–1906) purchased at auction by George A. Plimpton of New York City, in 1924, and presented to Harvard University. It is now reproduced for the first time.

Photograph by Whipple and Black, Boston, in 1858, in the class of 1829 album (in ms.), Harvard University Library.

Steel engraving, *The Mathematical Monthly*, ed. by J. D. Runkle, vol. 2, July 1860, facing page 329. Engraved by H. W. Smith from a daguerreotype by Southworth and Hawes in 1860, a "most accurate likeness." The same engraving appeared as frontispiece to *Annual of Scientific Discovery . . . for 1870*, Boston, 1870.

Very interesting reproduction of a photograph, full-length, of Peirce working at the blackboard and apparently taken about 1865; *Centennial Celebration of the United States Coast and Geodetic Survey*, Washington, 1916, p. 153. Copied in *The Scientific Monthly*, vol. 3, December, 1916, p. 618. Also on a plate opposite page 96 of *The Early Years of the Saturday Club 1855–1870*, by E. W. Emerson, Boston, 1918. Also now reproduced.

Photograph taken about 1872, nearly full-length, and reproduced in *The Harvard Book* by F. O. Vaille and H. A. Clarke, vol. 1, Cambridge, 1875, opposite page 172. Copied in *The Outlook*, New York, vol. 59, 1898, p. 323. Also in *Universities and Their Sons*, vol. 2, Boston, 1899, p. 228.

Reproduction of photograph, taken about 1872, F. C. Peirce, *Peirce Genealogy*, Worcester, 1880, oppo. p. 118.

Reproduction of photograph by G. W. Pach, New York, 1879, in *Harper's New Monthly Magazine*, March 1879, vol. 58, p. 508. Also in this MONTHLY, *1895*, oppo. p. 173.

Photograph, taken about 1879, reproduced as a steel engraving frontispiece in B. Peirce, *Ideality in the Physical Sciences*, 1881. Also in *Amer. Jl. Mathematics*, vol. 24, 1902, frontispiece. Woodcut of the same, apparently, in

The Harvard Register, May 1880, vol. 1, p. 91; in *Popular Science Mo.*, vol. 18, 1881, oppo. p. 578; and in M. King, *Benjamin Peirce . . . A Memorial Collection*, 1881.

Many other unpublished photographs of Benjamin Peirce are in Mr. Ellis's possession. There are also at least two other paintings of Peirce owned by grandchildren, one an inferior portrait by Wite and the other by a Miss Whitney from a photograph.

VI. The Writings of Peirce.

By R. C. Archibald.

[Solutions of problems.]
  The Mathematical Diary, New York, vol. 1, 1825, ed. by R. Adrian, pp. 281, 286.

[Solutions of problems, and a problem for solution.]
  The Mathematical Diary, New York, vol. 2, ed. by J. Ryan; no. X, 1828, p. 89; no. XI, 1830, pp. 116, 118; no. XIII, 1832, pp. 211–212, 216, 237, 244, 246, 310.

Laplace, *Mécanique Céleste translated with a commentary* by Nathaniel Bowditch.
  Boston, 4 volumes, 1829–1839.
  B. Peirce revised the entire work and detected many errata (compare the memoir in volume 4, pp. 61 and 140).

The American Almanac and Repository of Useful Knowledge. [For the years 1830–1851, the astronomical department was under the direction of B. Peirce.]
  Boston, 1829–1850.

On perfect numbers.
  The Mathematical Diary, vol. 2, no. XIII, 1832, pp. 267–277.
  Euler showed that every even perfect number is expressible in the Euclidean form $2^{n-1} \cdot p$ where $p = 2^n - 1$ is a prime. In Peirce's paper it is shown that there can be no odd perfect number "included in the form a^r, $a^r b^s$, $a^r b^s c^t$, where a, b, and c are prime numbers and greater than unity." In his *History of the Theory of Numbers*, vols. 1–3, 1919–1923, L. E. Dickson does not mention this paper. He does record that in 1844 "V. A. Lebesgue stated that he had a proof that there is no odd perfect number with fewer than four distinct prime factors." We now see that an American mathematician published a proof of this theorem twelve years earlier.

An elementary Treatise on Plane Trigonometry with its applications to heights and distances, navigation and surveying.
  Cambridge and Boston, Munroe, 1835, 7 + 90 pp. + 1 fold. pl.

First part of an elementary Treatise on Spherical Trigonometry.
  Boston, J. Munroe & Co., 1836, 4 + 71 pp. + 1 fold. pl.

An elementary Treatise on Sound: being the second volume of a course of natural philosophy designed for the use of high schools and colleges. Compiled by Benjamin Peirce.
  Boston, J. Munroe & Co., 1836, 56 + 220 pp.; diagrs. on 10 fold. pl.
  "The Catalogue of works relating to sound," pp. v–lvi, is interesting, and the list of titles connected with musical matters, pp. xvii–xlviii, valuable. The work is based on J. F. W. Herschel's treatise on sound, in the *Encycl. Metropolitana*. Reviewed in *The New York Review*, vol. 4, 1839, pp. 164–176.

[Problems and solutions.]
  The Mathematical Miscellany, ed. by C. Gill, New York, no. II: 1836, pp. 81–94, 94–97, 101–107; III: 1837, pp. 160–163 (194); IV: 1837, pp. 210–211, 233–234, 251–255, 258; V: 1838, pp. 289–290, 296, 309–311, 316–318, 327; VI: 1838, 359, 362–363, 383–387, 392–395, 397, 399; vol. 2, no. VII: 1839, pp. 16, 33–34, 42–47, 61, 63; VIII: 1839, 91, 92, 97–98, 110–113, 114, 117.
      For the question discussed vol. 2, pp. 97–98, compare *Sphinx Œdipe*, vol. 8, 1913, pp. 93–94. It seems probable that many problems proposed by "P." in the *Mathematical Miscellany* were due to Peirce. See, for example, vol. 1, pp. 53, 55, 109, 110, 257–258, etc.

BENJAMIN PEIRCE, 1865 (?)

An elementary Treatise on Algebra: to which are added exponential equations and logarithms.
    Boston, J. Munroe & Co., 1837, 10 + 276 pp.
    Another edition, 1842, 4 + 284 pp.; 1843; fifth ed., 1845; sixth ed., 1846; other editions or
        reprints, 1850, 1851, 1855, 1858, 1860; and published by W. H. Dennett in 1864, 1865,
        1870.

An elementary Treatise on Plane and Solid Geometry.
    Boston, J. Munroe & Co., 1837, 20 + 159 pp. + 6 fold. pls.
    Other editions or reprints, 20 + 3-150 p. + 6 fold. pls., 1841, 1847, 1851, 1853, 1855,
        1857, 1860; and publ. by W. H. Dennett in 1863, 1865, 1866, 1867, 1868, 1869, 1870,
        1871, 1872.

An account of Mr. Talbot's "Researches in the integral calculus" [*Philosophical Transactions,*
    vols. 54–55, 1836–37].
    The Mathematical Miscellany, ed. by C. Gill, vol. 1, no. VI, 1838, pp. 404–411.

[Anonymous review of Laplace's *Mécanique Céleste,* vols. 1, 3–5; N. Bowditch's translation, vols.
    1–4; and A. Young, D. A. White, and J. Pickering on Bowditch.]
    N. Amer. Rev., vol. 44, 1839, pp. 143–180.

[Anonymous review of J. Pickering, D. A. White, and A. Young on N. Bowditch.]
    The New York Review, vol. 4, 1839, pp. 308–323.

An elementary Treatise on Plane Geometry. . . . Printed for the use of the blind.
    Boston, At the Press of the Perkins Institution and Massachusetts Asylum, 1840. 71 pages
    of definitions and demonstrations + 19 pages embossed diagrams.
        In Boston line type; size 11 x 10 inches.

An elementary Treatise on Plane and Spherical Trigonometry, with their application to navigation,
surveying, heights and distances and spherical astronomy, and particularly adapted to explaining
the construction of Bowditch's Navigator and the Nautical Almanac.
    Boston, J. Munroe & Co., 1840, 4 + 428 pp. + 5 fold. pls.
    Third edition with additions, 1845, 4 + 449 pp. + 4 fold. pls.
    New edition revised, with additions, 1852, 6 + 360 pp. + 5 fold. pls.
    Another edition or reprint, 1861.
    Pages 317–357 + plates 4–5 were reprinted in 1852 with the following title page: "*The
        chapter on Eclipses extracted from Peirce's Spherical Astronomy. For the Use of the
        Nautical Almanac.*

An elementary Treatise on Curves, Functions, and Forces. Volume first; containing analytical
geometry and the differential calculus, 1841: Volume second; containing calculus of imaginary quan-
tities, residual calculus, and integral calculus, 1846.
    Boston, James Munroe & Co., 8 + 304 pp. + 14 pls. + 8 + 290 pp. + fold. pl.
    New edition of volume 1, 1852. 7 + 301 pp. + 14 pls.
    A third volume of this work, dealing with the applications of analytic mechanics, was
        projected, but not published. Probably the treatise issued in 1855 took its place.
        The following text by J. M. Peirce is based on this treatise: *A Text-book of Analytic
        Geometry; on the basis of Professor Peirce's Treatise.* Cambridge, 1857, 7 + 228 pp.
        + 6 fold. pl. Extracts from the preface: "I would acknowledge, in closing, my
        obligations for the aid and encouragement which I have received from others.
        Professor Peirce has given me the benefit of his advice in repeated instances. What-
        ever merit the book may have is owing, in a great degree, to the assistance of Mr.
        C. W. Eliot, who, besides many less definite, but important services, has read and
        criticised a considerable part of the manuscript before it was sent to the press."
        In 1845 the following work was published in Boston by Thomas Hill, afterwards
        president of Harvard College: *An Elementary treatise on Arithmetic, designed as an
        Introduction to Peirce's Course of Pure Mathematics* [that is the series of works
        published 1835–1846] *and as a sequel to the Arithmetics used in the High Schools of
        New England.*

The Cambridge Miscellany of Mathematics, Physics, and Astronomy, no. I (April, 1842), edited by
    B. Peirce; no. II (July, 1842), no. III (October), no. IV (last, January, 1843), edited by B.
    Peirce and J. Lovering.
    Boston, James Munroe & Co., 1–48 + 49–96 + 97–144 + 145–192 pp. + 3 plates.

Contributions by B. Peirce: Problems and Solutions, pp. 23–24, 58, 60–61, 66–72, 97, 102, 119, 145, 149, 155, 156, 159, 168; "American astronomical and magnetic observers," pp. 25–28; "Distances of the fixed stars," pp. 28–31; "Meteors," pp. 44–46; "Varieties of climate," pp. 46–47; "The barometer," p. 48; "On Espy's theory of storms," pp. 141–144. Joseph Lovering (1813–92) was teacher of mathematics and natural philosophy at Harvard for the fifty years prior to 1888.

[Anonymous review of S. C. Walker's work on Meteors.]
N. Amer. Rev., vol. 56, 1843, pp. 409–435.

On the perturbation of meteors approaching the earth.
Amer. Philos. Soc., Trans. n.s., vol. 8, 1843, pp. 83–86.
A letter dated Dec. 24, 1840, to S. C. Walker. Presented to the Society, January 15, 1841.

Bowditch's Useful Tables. [Preface by J. I. Bowditch, pages iii–v; "Remarks of Professor Pierce" (*sic*), pp. vii–viii.]
New York, E. & G. Blunt, 1844. Other editions were published in 1856, 1859, 1863, and 1866. An edition was issued by the U. S. Bureau of Navigation in 1885. Recent editions (such as the one for 1911) do not contain Peirce's remarks.
The tables in question are taken from Nathaniel Bowditch's *Practical Navigator*.

[Elements of the] third comet of 1845.
Amer. Jl. Sci., vol. 49, 1845, pp. 220–221.

The latitude of Cambridge Observatory, in Massachusetts, determined, from transits of stars over the prime vertical, observed during the months of December, 1844, and January, 1845, by William C. Bond, James D. Graham, and George P. Bond.
Amer. Acad., Memoirs, Boston, n.s., vol. 2, 1846, pp. 183–203.

[Orbits for Bond's comet.]
Boston Courier, March 27, 1846, p. 2, col. 2.
Also in *Amer. Jl. Sci.*, n.s., vol. 1, 1846, p. 348.
Letter dated March 26, 1846.

The Perturbations of Uranus.
Boston Courier, April 30, 1847, p. 2, col. 3.
Also in *Amer. Acad., Proc.*, vol. 1 (1846–48), 1848, pp. 144–145.
Also in *Amer. Jl. Sci.*, n.s., vol. 4, 1847, pp. 132–133.
Communication dated April 29, 1847.

Mass of Neptune.
Boston Courier, October 25, 1847, p. 2, cols. 2–3.
Also in *National Intelligencer*, Washington, October 26, 1847, p. 2, col. 5.
Leverrier's reply is in the *Intelligencer*, March 10, 1848, p. 2, cols. 5–6, and in the *Boston Daily Advertiser*, March 20, p. 1, cols. 5–6. To this Peirce replied in a letter dated March 13, 1848; this is in the *Intelligencer*, March 23, p. 1, col. 2; and in the *Advertiser*, March 24, p. 2, col. 2, "Leverrier and Mr. Peirce." The following letter from Asa Gray, the distinguished botanist then professor of natural history at Harvard, is dated March 26:
Dear Peirce
When I read, in the Daily, your letter, I was on the point of sitting down to write you a line to tell you that I think it a *perfect gem*, and the most beautiful contrast to the *Johnny Crapeau* vociferation of Le Verrier. I am perfectly charmed with its spirit, and all that I have heard speak of it have taken the same view. I am not alone, therefore, in the opinion that it does you the highest credit and it is just the style of reply calculated to place you at the greatest advantage. As one zealous for the highest interests and character of American Science and American *Savans*, I thank you most sincerely, and am
Faithfully yours,
A. Gray
Also in *The Siderial Messenger*, Cincinnati, vol. 2, pp. 28–29, 1847.
Letter dated October 22, 1847.

[Elements of an elliptic orbit of De Vico's fourth comet.]
Amer. Acad., Proc., vol. 1 (1846–48), 1848, pp. 39–42.

Notice of the computations of Mr. Sears C. Walker, who found that a star was missing in the *Histoire Céleste Française*, observed by Lalande on the 10th of May, 1795, near the path of the planet Neptune at that date, which may possibly have been this planet.

> *Amer. Acad., Proc.*, vol. 1 (1846–48), 1848, pp. 57–68   See also pp. 41–42.
> Pages 57–65 appeared also in *The Siderial Messenger*, Cincinnati, vol. 1, pp. 125–128, 1847, under the title "The planet Neptune"; some parts of the other pages are given, in substance, on pages 85–86 of the same volume in an article entitled "Le Verrier's planet." Pages 64–67 appeared as a quotation in *Amer. Jl. Sci.*, n.s., vol. 3, 1847, pp. 441–443.

[Review of Nichol's *Contemplations on the Solar System*.]

> *N. Amer. Rev.*, vol. 66, 1848, pp. 253–255.

[Formulæ for the perturbations of Neptune's longitude and radius vector.]

> *Amer. Acad., Proc.*, vol. 1 (1846–48), 1848, pp. 285–295.

[Investigations into the action of Neptune upon Uranus.]

> *Amer. Acad., Proc.*,[1] vol. 1 (1846–48), 1848, pp. 332–342.
> (a) Perturbations [of Neptune]; (b) Elements of Neptune; (c) Satellite of Neptune; (d) [Perturbations of Uranus].
> *Mo. Notices R. Astr. Soc.*, vol. 8 (1847–48), 1848, pp. 38–40, 128, 202–203.

Ueber die Störungen des Neptuns.

> *Astron. Nachrichten*, vol. 27, cols. 215–218, 1848.

[Calculations on the perturbations of Uranus.]

> *Amer. Jl. Sci.*, n.s., vol. 5, 1848, pp. 435–436.

Development of the perturbative function of planetary motion.

> *Astr. Jl.*, vol. 1, pp. 1–8, 1849; pp. 31–32, 33–36, 1850.

Certain methods of determining the number of real roots of equations applicable to transcendental as well as to algebraic equations [abstract].

> *Amer. Assoc. Adv. Sci., Proc.*, vol. 1 (1848), 1849, pp. 38–39.

A. Guyot, *The Earth and Man*. Translated from the French by C. C. Felton. Boston, 1849.

> Quotation from the preface: "Besides Prof. Felton, who read all the proof sheets, the author returns his acknowledgments to Professors Agassiz, Peirce and Gray who have had the goodness to revise portions of them."

On Fresnel's dioptric apparatus for lighthouses, by B. Peirce and J. Lovering.

> *Franklin Inst., Jl.*, s. 3, vol. 18, 1849, pp. 249–252.

"*Poor Richard.*"   *Poor Richard's Almanac for 1850(–52) as written by Benjamin Franklin for the years 1733, 1734, 1735, (1736–1741). The astronomical calculations by . . . Benj. Peirce.* New York, John Doggett, Jr., 1849(–51).

(a) On the connection of comets with the solar system; (b) On the relation between the elastic curve and the motion of the pendulum; (c) Mathematical investigation of the fractions which occur in phyllotaxis.

> *Amer. Assoc. Adv. Sci., Proc.*, vol. 2 (1849), 1850, pp. 118–122; 128–130; 444–447.
> As to the topic discussed in (c) compare my articles on Golden section and the Fibonacci series in this MONTHLY, *1918*, 232–238; or better in Jay Hambidge, *Dynamic Symmetry*, New Haven, 1920, pp. 152–157. See also C. Wright, "On the phyllotaxis," *Astron. Jl.*, vol. 5, pp. 22–24, 1856.

(a) Note of Professor Peirce to the editor [demonstrating the parallelogram of forces]; (b) On the orbit of α Virginis regarded as a double star.

> *Astr. Jl.*, vol. 1, pp. 23, 138–139, 1850.

On the constitution of Saturn's rings.

> *Amer. Jl. Sci.*, vol. 12, 1851, pp. 106–108.
> Also in *Astr. Jl.*, vol. 2, pp. 17–19, 1851.
> German translation: Ueber die Beschaffenheit des Saturnringes, *Annalen der Physik und Chemie*, vol. 160, 1851, pp. 313–319.

---

[1] There are numerous references in the *Proceedings* to papers read by Peirce but not published. See vol. 1 (1846–48), 1848, p. 185; vol. 2 (1848–52), 1852, pp. 111–2, 147, 235, 240, 250, 256, 258, 282, 289–90, 298, 310; vol. 3 (1852–57), 1857, pp. 8, 9, 28, 31, 67, 83.

Report of the Committee upon Prof. Mitchell's system of astronomical observations by Benjamin Peirce, chairman.
>*Amer. Assoc. Adv. Sci., Proc.*, vol. 5, 1851, pp. 69–71.

(a) [Report on the results of the U. S. Coast Survey, by B. Peirce for the Committee: B. Peirce, D. Treadwell, J. I. Bowditch, and J. Lovering]; (b) [Report by B. Peirce, E. N. Horford, and J. Lovering on a paper, entitled, "Description of the causes of the explosion of steam boilers, and of some newly-discovered properties of heat and other matters; for the purpose of showing that the application of steam for the production of motive force is susceptible both of immense improvement and economy," by James Frost]; (c) [On a new method of computing the constants of the perturbation function of planetary motion]; (d) Report by B. Peirce (in behalf of the committee D. Treadwell, B. Peirce, J. Lovering, H. L. Eustis, and M. Wyman) concerning C. H. Davis's paper on the deterioration of Boston harbor.
>*Amer. Acad., Proc.*, vol. 2 (1848–52), 1852, pp. 124–128; 129–130; 197–198; 288–289.

An account of Longstreth's lunar formula.
>*Amer. Assoc. Adv. Sci., Proc.*, vol. 6 (1851), 1852, pp. 143–144.

The American Ephemeris and Nautical Almanac for the year 1855 [vol. 1] (–1861).
>Washington, 1852(–1858).
>>From the preface: "The theoretical department of the work has been placed under the special direction of Professor Benjamin Peirce, LL.D., and most of the calculations have passed under his final revision."

(a) Note upon the conical pendulum; (b) Criterion for the rejection of doubtful observations.
>*Astr. Jl.*, vol. 2, 1852, pp. 137–149; 161–163.
>>An elucidation of the second of these papers is given in the paper on the Criterion, published by Peirce in 1878. See further the notes on this subject in Section V, p. 13.

The semidiameters of Venus and Mars investigated. From the observations made with the mural-circle of the Naval Observatory at Washington during the years 1845 and 1846.
>*Astr. Jl.*, vol. 3, pp. 9–10, 1852.

Tables of the Moon; constructed from Plana's theory, with Airy's and Longstreth's corrections, Hansen's two inequalities of long period arising from the action of Venus, and Hansen's values of the secular variations of the mean motion and of the motion of the perigee. Arranged in form designed by . . . Benjamin Peirce.
>Washington, For the use of the Nautical Almanac, 1853. 4to, 326 pp.
>See also *Tables of the Moon's Parallax*, 1856.
>>The second edition (1865, 348 pp.) was "the same as the first, with the exception of the correction of typographical errors; the substitution of the Tables of the Moon's Parallax constructed from Walker's and Adams' formulas, in the place of the original Parallax Tables; and the addition of a Table adapted to a convenient modification of the method of computing the latitude, by Professor J. D. Runkle.
>>"A third edition will shortly be issued, of which the basis will still be Plana's theory, while the Tables will be corrected to conform to the new Solar Parallax, and the corrected elements of the Moon's orbit." (Preface; the third edition does not seem to have been published.)
>>The work was reviewed in *Mo. Notices R. Astr. Soc.*, vol. 14, pp. 26–32, 184. On page 32 it is remarked that "the arrangement . . . is the result of a plan devised by Professor Peirce. It is very clear and masterly, and is in every respect worthy of that eminent mathematician."

On longitudes from moon culminations.
>*Coast Survey, Report for 1853*, Washington, 1853, app. 31, p. 84.

Address of Professor Benjamin Peirce, president of the American Association for the year 1853, on retiring from the duties of president. [Printed by order of the Association.]
>[Cambridge], 1854, 17 p.
>Also in *Amer. Assoc. Adv. Sci., Proc.*, vol. 8 (1854), 1855, pp. 1–17.
>Also bound in limp cloth, gilt edges, with cover title *The Song of Geometry*, and with special title-page and dedication. The copy of the title page is as follows: *Ben Yamen's Song of Geometry, sung by the Florentine Academy, at the accession of Her Majesty the Queen, degraded into prose by Benjamin the Florentine*, Cloverden, 1854.
>Also most of pages 2–5 appears on pages 105–108 of *The Early Years of the Saturday Club, 1855–1870*, by E. W. Emerson, Boston, 1918.

Residual differences between the theoretical and observed longitudes of Uranus, from the theories of Peirce, LeVerrier and Adams.
> *Amer. Phil. Soc., Proc.*, vol. 5, 1854, p. 16.

(a) Elements of the comet, 1854, III; (b) Quantities to be added to the solar ephemeris of the American Nautical Almanac to obtain that given by Hansen's solar tables with the obliquity of the ecliptic of the Nautical Almanac; (c) The investigation of the catenary upon a cone of revolution with a vertical axis.
> *Astr. Jl.*, vol. 4, pp. 7, 9, 27–29, 1854.

Report upon the determination of longitude by moon-culminations.
> *Coast Survey, Report for 1854*, Washington, 1854, app. 36, pp. 108–120.

On the Adams prize-problem for 1856.
> *Astr. Jl.*, vol. 4, pp. 27–29, 1854.
>> At the conclusion of the article are the words "to be continued"; no continuation has been found.

[Report of the Committee of the American Academy of Arts and Sciences on a Program for organization of the Smithsonian Institution, December 8, 1847.]
> *Smithsonian Institution, Eighth Annual Report*, Washington, 1854, pp. 148–155.
> Also in *The Smithsonian Institution. Documents Relative to its Origin and History (Smithsonian Misc. Colls.*, vol. 17, Washington, 1879, pp. 964–970).
> The Committee consisted of E. Everett (Chairman), Jared Sparks, Benjamin Peirce, H. W. Longfellow, and Asa Gray.

Physical and Celestial Mechanics . . . developed in four Systems of Analytic Mechanics, Celestial Mechanics, Potential Physics, and Analytic Morphology. Then the second title page: *A System of Analytic Mechanics.*
> Boston, Little, Brown & Co., 1855, 39 + 496 pp. + a fold. pl.
>> This work on mechanics was intended as the first of a series of four volumes, the other three to be respectively on Celestial Mechanics, Potential Physics, and Analytic Morphology.
>> Extract from preface: "I have . . . reexamined the memoirs of the great geometers, and have striven to consolidate their latest researches and their most exalted forms of thought into a consistent and uniform treatise. If I have, hereby, succeeded in opening to the students of my country a readier access to these choice jewels of intellect, if their brilliancy is not impaired in this attempt to reset them, if in their new constellation they illustrate each other and concentrate a stronger light upon the names of their discoverers, and still more, if any gem which I may have presumed to add is not wholly lustreless in the collection, I shall feel that my work has not been in vain. The treatise is not, however, designed to be a mere compilation. The attempt has been made to carry back the fundamental principles of the science to a more profound and central origin; and thence to shorten the path to the most fruitful forms of research. See further comments in Section V of this monograph.
>> Reviewed in: *Christian Examiner*, Boston, vol. 64, 1858, pp. 276–293 [by Thos. Hill]; *N. Amer. Rev.*, vol. 87, 1858, pp. 1–21 [by T. Hill].
> Another issue with new title-page, New York, Van Nostrand, 1865; also 1872.

On the method of determining longitudes by occultations of the Pleiades.
> *Coast Survey, Report for 1855*, Washington, 1855, app. 42, pp. 267–274.

Six articles upon the Smithsonian Institution . . . together with the letters of Professor Peirce and Agassiz.
> *Boston Post*, January 27, February 5, 7, 13, 21, 22, 1855.
> Also as a pamphlet, Boston, Printed at the office of the *Boston Post*, 1855, 44 pp.
>> The letter of Professor Peirce was dated January 29, 1855. The articles were signed "N. P. D." The last three of the five paragraphs of Peirce's letter were published in *The Congressional Globe, Appendix* (33d Congress, 1853–55, House of Representatives, February 27, 1855), vol. 31, p. 285, and reprinted in *The Smithsonian Institution, Documents relative to its Origin and History (Smithsonian Miscellaneous Collections*, vol. 17), Washington, 1879, pp. 588–9, 619.

Letter of Professor Peirce to President Quincy, with two letters from Admiral Beaufort annexed to it, and a list of zenith stars from Professor Airy.
>   *Annals of the Harvard Observatory*, vol. 1, 1856, pp. xciv–xcv.
>   The letter was dated May 10, 1845.

Opening address of Professor Benjamin Peirce . . . President of the Association.
>   *Amer. Assoc. Adv. Sci., Proc.*, vol. 7 (1853), 1856, pp. xvii–xx.

(a) Abstract of a paper on researches in analytic morphology. Transformation of curves; (b) Abstract of a paper upon the solution of the Adams prize problem for 1867; (c) Abstract of a paper on partial multipliers of differential equations; (d) Abstract of a paper upon the catenary on the vertical right cone; (e) Abstract of a paper upon the motion of a heavy body on the circumference of a circle which rotates uniformly about a vertical axis; (f) Abstract of a paper on the resistance to the motion of the pendulum; (g) Method of determining longitudes by occultations of the Pleiades.
>   *Amer. Assoc. Adv. Sci., Proc.*, vol. 9 (1855), 1856, pp. 67–74, 97–102.

Tables of the Moon's Parallax, constructed from Walker's and Adams's formulæ, arranged as a supplement to the first edition of Peirce's Tables of the moon.
>   Washington, For the U. S. Nautical Almanac Office, 1856, pp. 303–329.

Working plan for the Foundation of a University.
>   Cambridge, Mass., 1856, 4 pp.
>   "Printed for private and confidential circulation among the advocates and patrons of the University" [Harvard].

On the determination of longitude by occultations of the Pleiades.
>   *Coast Survey, Rept. for 1856*, Washington, 1856, app. 24, pp. 191–197.

An investigation of the cases of complete solution by integration by quadratures of the problem of the motion of a material point acted upon by forces which emanate from a fixed axis.
>   *Astr. Jl.*, vol. 5, pp. 38–39, 1857.

[Report, dated June 29, 1857, by a committee consisting of Benjamin Peirce (chairman), Louis Agassiz, B. A. Gould, and E. N. Horsford, on spiritualistic phenomena presented by a Dr. Gardner in an attempt to win a $500 prize offered in 1857 by the *Boston Courier*.]
>   *Boston Courier*, July 1, 1857, p. 2, col. 2.
>   Also in Epes Sargent, *Planchette; or the Despair of Science*, Boston, 1869, pp. 10–11; see also p. 13. See further, "The Cambridge professors" in T. L. Nichols, *A Biography of the Brothers Davenport*, London, 1864, pp. 83–91; and G. A. Redman, *Mystic Hours; or Spiritual Experiences*, New York, 1859, pp. 307–317.

Determination of longitudes by occultations of the Pleiades and solar eclipses.
>   *Coast Survey, Rept. for 1857*, Washington, 1857, app. 29, pp. 311–314.

(a) Note on the Red Hill catalogue of circumpolar stars; (b) Note on the extension of Lagrange's theorem for the development of functions.
>   *Astr. Jl.*, vol. 5, pp. 137, 164, 1858.

On the formation of continents.
>   *Canadian Jl. of Industry, Science, and Art*, Canadian Institute, Toronto, n.s., vol. 3, 1858, pp. 69–70.

(a) Problem; (b) Propositions on the distribution of points on a line; (c) Note on two symbols.
>   *Mathematical Monthly*, ed. by J. D. Runkle, vol. 1, pp. 11, 16–18, 58, 60, 1858; 167–168, 170, 1859.

Cotidal lines of an inclosed sea, derived from the equilibrium theory.
>   *Coast Survey, Rept. for 1858*, Washington, 1858, app. 30, pp. 210–213.

Defence of Dr. Gould by the Scientific Council of the Dudley Observatory [Albany, N. Y.].
>   Albany, Weed, Parsons & Co., 1858, 93 pp.
>   Signed by Joseph Henry, A. D. Bache, Benjamin Peirce, Dudley Observatory, July 1858. See also *The Dudley Observatory and the Scientific Council, Statement of the Trustees*, Albany, 1858; letters of Peirce, pp. 54, 101, 102. Also, *A Key to the "Trustee's Statement." Letters to the Majority of the Trustees of the Dudley Observatory, showing the Misrepresentation, Garblings, Perversions of their Misstatements*, by George H. Thacher. Albany, Atlas & Argus, Oct. 1858, p. 126.
>   Third edition, Albany, 1858.

On the theory of the comet's tail.
Astr. Jl., vol. 5, pp. 186–188, 1858; vol. 6, pp. 50–56, 1859.

[Resolutions by B. Peirce, at meeting of A. A. A. S., in Springfield, Mass., voting thanks to ladies of Springfield.]
The Atlas and Daily Bee, Boston, vol. 34, Aug. 11, 1859, p. 1, col. 7.

[W. C. Bond, director of Harvard Observatory; obituary notice.]
Amer. Acad., Proc., vol. 4 (1857–60), 1860, pp. 163–166.

Lettre addressée à M. le président de l'Académie des Sciences sur la constitution physique des comètes.
Comptes Rendus de l'Académ. d. Sc., vol. 51, 1860, pp. 174–176.

(a) [Abstract of a memoir on the peculiarities of astronomical observers;] (b) Memoir upon the tail of Donati's Comet.
Amer. Acad., Proc., vol. 4 (1857–60), 1860, pp. 197–199, 202–206.

Cyclic Solutions of the school-girl puzzle.
Astr. Jl., vol. 6, pp. 169–174, 1860.
> The problem here discussed is the following: "A given number, f, of girls are required to walk in a given number, g, of ranks, of which each rank consists of a given number, k, of girls; subject to the condition that each girl is to walk once, and only once, in the same rank with every other girl." This important paper is apparently one of a series inspired by T. P. Kirkman's problem published in the *Lady's and Gentleman's Diary* for 1850, p. 48: "Fifteen young ladies in a school walk out thrice abreast for seven days in succession; it is required to arrange them daily, so that no two shall walk twice abreast." An account of this problem and some references to the literature are given in W. W. R. Ball, *Mathematical Recreations*, tenth edition, London, 1922, pp. 193–223; see also *Messenger of Mathematics*, vol. 41, 1911, pp. 33–56; *Jahrbuch über die Fortschritte der Mathematik*, 1911, p. 250, and W. Ahrens, *Mathematische Unterhaltungen und Spiele*, vol. 2, Leipzig, 1919, pp. 102–117.

(a) Report upon the determination of the longitude of America and Europe from the solar eclipse of July 28, 1851; (b) Report on an example for the determination of longitudes by occultations of the Pleiades.
Coast Survey, Rept. for 1861, Washington, 1861, (a) app. 16, pp. 182–195; (b) app. 17, pp. 196–221.

(a) Abstract of a memoir upon the attraction of Saturn's ring; (b) Upon the system of Saturn.
Amer. Acad., Proc., vol. 5 (1860–62), 1862, pp. (a) 353–354; (b) 379–380.

(a) On the computations of the occultations of the Pleiades for longitude; (b) Upon the tables of the moon used in the reduction of the Pleiades.
Coast Survey, Rept. for 1862, Washington, 1862; (a) app. 12, pp. 155, 156; (b) app. 13, pp. 157, 158.

Report upon the occultations of the Pleiades in 1841–42.
Coast Survey, Rept. for 1863, Washington, 1863, app. 17, pp. 146–154.

On the computations for longitudes by occultations of the Pleiades.
Coast Survey, Rept. for 1864, Washington, 1864, app. 11, p. 114.

(a) Report on the progress of determining longitude from occultations of the Pleiades [continued, compare *Report for 1863*]; (b) Method of determining the corrections of lunar semi diameter, mean place, ellipticity of orbit, longitude of perihelion, coëfficient of annual parallax, and longitude of Europe and America from the occultation of the Pleiades.
Coast Survey, Rept. for 1865, Washington, 1865; (a) app. 12, pp. 138–146; (b) app. 13, pp. 146–149.

On the lunar bolis.
Amer. Acad., Proc., vol. 6 (1862–65), 1866, p. 36.

The Saturnian system.
Nat. Acad. Sci., Mem., vol. 1, 1866, pp. 263–286.

Coast Survey, Report, B. Peirce, Superintendent [For the years 1867–1873].
  House Executive Documents, Washington, 1867–1873.
      Details concerning the exact number of these reports in the *Documents* and the number
        of pages in each report may be found in B. P. Poore, *A Descriptive Catalogue of the
        Government Publications of the United States, 1774–1881,* Washington, 1885.   See also
        E. L. Burchard, *List and Catalogue of The Publications issued by the U. S. Coast and
        Geodetic Survey, 1816–1902,* Washington, 1908.

Obituary on Alexander Bache.
  Coast Survey, Rept. for 1867, Washington, 1867, app. 19, p. 330.

Communication of vibration.
  Amer. Assoc. Adv. Sci., Proc., vol. 16 (1867), 1868, pp. 17–18.

Report on Weights and Measures.
  Washington, Coast Survey, 1869, 4 pp.
      Report upon the progress made in the construction of metric standards of length, weight,
        and capacity, in pursuance of a joint resolution of Congress of July 27, 1866.

The solar eclipse of December 22, 1870.
  Coast Survey, Rept. for 1870, Washington, 1870, app. 16, pp. 229–232.

Linear Associative Algebra (Lithographed).
  Washington City, 1870, 153 pp.
      Edition limited to 100 copies issued through "labors of love" by persons engaged on
        the Coast Survey.   This work was developed from papers read before the National
        Academy of Sciences, 1866–1870.
  New edition, with addenda and notes by C. S. Peirce, son of the author.
      Amer. Jl. Math., vol. 4, 1881, pp. 97–229.
      Reprinted, New York, Van Nostrand, 1882, 4 + 133 pp.
          This contains pp. 120–125, a reprint of: (1) his article on the uses and transformations
            of linear algebra, published in 1875; and (2) C. S. Peirce's notes, pp. 125–
            133, which appeared at the same time.
              On page 656, volume 3 (1869) of R. P. Graves's *Life of Sir William
            Rowan Hamilton,* occurs the following with reference to Peirce's work as it
            appeared in the *American Journal:* "The author of this Paper in a *note,* on p.
            105, makes objection to Quaternions on the ground of the treatment of imagi-
            naries.   A reply to this objection may, I believe, be gathered from what will
            be found stated by Sir William Hamilton in pages 578, 579 of vol. II and
            pages 84, 85 of vol. III of this work, as well as *passim* in the correspondence
            with Professor De Morgan."
              See notes in Section V, pp. 15–16.

[Problem proposed.]   Given the skill of two billiard players at the three-ball game, to find the
  chance of the better player gaining the victory if he gives the other a *grand discount.*
  Our Schoolday Visitor, Philadelphia, vol. 15, 1871, p. 220, problem 108.
  Also as problem 71 in *The Mathematical Visitor,* vol. 1, p. 46, 1878; solution by the proposer
      on p. 69, 1879.   Compare Peirce's paper, "Probabilities at the three-ball game of
      billiards," 1877.

Observations of the eclipse of December 22, 1870, at Catania.
  Boston Daily Advertiser,   , 1871.
  Also in *Amer. Jl. Sci.,* s. 3, vol. 1, 1871, p. 155*.
      Letter dated Catania, Dec. 22, 1870.

On the mean motions of the four outer planets.
  Amer. Jl. Sci., s. 3, vol. 3, 1872, pp. 67–68.
      From a letter to H. A. Newton, dated December 13, 1871.

*Harbor of New York: its Condition, May, 1873.   Letter . . . to the Chamber of Commerce of New
  York, with the report of Prof. Henry Mitchell on the Physical Survey of the Harbor.*
  New York, Press of Chamber of Commerce, 1873.   3–38 pp. + 8 charts and tables.
      The letter occupies pages 3–5.

[On the formation of the shell of the earth by shrinkage.]
  Amer. Acad., Proc., vol. 8 (1868–73), 1873, pp. 106–108.

BENJAMIN PEIRCE, 1879

The rotation of the planets as a result of the nebular theory.
Nature, vol. 8, 1873, pp. 392–393.
Reprint of a report of the 1873 meeting of the A. A. A. S., in the *New York Tribune*.
Ocean lanes for steamships.
Amer. Acad., Proc., vol. 9 (1873–74), 1874, pp. 228–230.
On the uses and transformations of linear algebras.
Amer. Acad., Proc., vol. 10, 1875, pp. 395–400.
Also in *Amer. Jl. Math.*, vol. 4, 1881, pp. 216–221; in this form reprinted in *Linear Associative Algebra*, 1882, pp. 120–125.
A new system of binary arithmetic.
Coast Survey, Rept. for 1876, Washington, 1876, app. 6, pp. 81–82.
The conflict between science and religion.
Unitarian Review, Boston, vol. 7, 1877, pp. 656–666.
Also reprinted with cover title, Boston, 1877, 12 pp.
A discourse delivered in the First Church, Boston, May 6, 1877.
Qualitative algebra.
Johnson's New Universal Cyclopædia, New York, vol. 3, 1877, pp. 1487–88.
(a) Probabilities at the three-ball game of billiards; (b) on Peirce's criterion.
Amer. Acad., Proc., vol. 13 (1877–1878), 1878, pp. (a) 141–144; (b) 348–351.
The second of these papers was in elucidation of the paper on the same subject published in 1852. Compare page 13 of this monograph.
The National Importance of Social Science in the United States. An address delivered by Professor Benjamin Peirce, at the opening of the session of the American Social Science Association at Cincinnati, 18 May, 1878.
Boston, Little, Brown & Co., 1878, cover-title, 16 p.
Also in *Journal of Social Science*, no. 12, 1880, pp. xii–xxi.
[Problem proposed] 5564. Find the probabilities at a game of a given number of points, which is played in such a way that there is only one person who is the actual player, and when the player is successful he counts a point, but when he is unsuccessful, he loses all the points he has made and adds one to his opponent's score.
Educational Times, London, vol. 31, 1878, p. 88; solution, pp. 135–136.
Also in *Mathematical Questions with Solutions from the Educational Times*, vol. 29, 1878, pp. 72–73.
Also as problem 66 in *The Mathematical Visitor*, Erie, Pa., vol. 1, p. 45, 1878; Peirce's solution is given on page 66, 1879.
Also in E. J. Boudin, *Leçons de Calcul des Probabilités*, edited by P. Mansion, Paris, 1916, pp. VIII and 36 ff. Several solutions of the problem are given, one by A. Claeys, another by A. Demoulin, and its connections with important theory are set forth.
[Problem proposed] 5968. If two bodies revolve about a centre, acted upon by a force proportional to the distance from the centre, and independent of the mass of the attracted body, prove that each will appear to the other to move in a plane, whatever be the mutual attraction.
Educational Times, vol. 32, 1879, p. 152; solutions, vol. 33, 1880, pp. 141 (by C. J. Munro) and 309 (by Asaph Hall).
Also in *Mathematical Questions . . .* , vol. 33, 1880, p. 91; vol. 34, 1881, p. 111.
Also as problem 145 in *The Mathematical Visitor*, vol. 1, p. 84, 1879; quaternion solution by the proposer, and a solution by De V Wood, p. 146, 1880.
Internal constitution of the earth.
Coast Survey, Rept. for 1879, Washington, 1879, app. 14, p. 201.
[Problems proposed] (a) 174. To find by quadratic equations a triangle of which the angles are given and the distances of the vertices from a given point in the plane of the triangle. (b) 202. Find a curve which is similar to its own evolute.
The Mathematical Visitor, vol. 1, pp. 99, 116, 1880; solutions of 174 by W. Hoover and W. Siverly, p. 174, 1881; solution of 202 by A. S. Christie, vol. 2, pp. 21–2, 1882.
Propositions in cosmical physics.
Amer. Acad., Proc., vol. 15, 1880, p. 201.

The intellectual organization of Harvard University.
The Harvard Register, April 1880, vol. 1, p. 77.

Ideality of the Physical Sciences. Edited by J. M. Peirce.
Boston, Little, Brown & Co., 1881, 7 + 9–211 pp. Portrait frontispiece (fine steel engraving).
The editing consisted in verbal changes and the addition of footnotes and the appendix (pp. 195–211).

Reprint, 1883.
This volume contains the six lectures delivered by B. Peirce in February and March, 1879, in a Lowell Institute course, Boston, Mass. They were also given, January 20 to February 5, 1880, at the Peabody Institute, Baltimore. The lectures are entitled: 1. Ideality in Science; 2. Cosmogony; 3. From nebula to star; 4. Planet, comet, and meteor; 5. The cooling of the earth and the sun; 6. Potentiality. The dedication is as follows:

I dedicate
these lectures
To my wife
with my whole heart.

Benjamin Peirce.

Cambridge, 790320

The last of these lectures was quoted under the heading, "Prof. Peirce on the spiritual body" in *Banner of Light*, Boston, May 3, 1879.

ADDENDA

¶ To the list of references given in footnote 1, page 9, and footnote 1, page 11, references may be given to J. E. Hilgard's obituary notice of Peirce in *Report of the Superintendent of the U. S. Coast and Geodetic Survey for the year ending June, 1881*, Washington, 1883, pp. 8–9; and to E. S. Holden, *Memorials of William Cranch Bond, director of the Harvard College Observatory 1840–1859, and of his son George Phillips Bond, director of the Harvard College Observatory 1859–1865*, San Francisco and New York, 1897. In this latter work are numerous references to Peirce in the index, some of them to material of special interest in portraying his character (pp. 26 and 163), and in exhibiting Leverrier's reaction (pp. 91–92) to Peirce's criticisms of his discovery of Neptune (cf. pp. 14 and 22 of this Monograph); Gauss's comments made in this connection in 1851 are recorded on page 109 where reference is also made to Gauss's two sons and a grandson in St. Louis. Peirce's notice of W. C. Bond (compare p. 27 of this Monograph) is reprinted on pages 43–46 of this work.

¶ In addition to the references to "Peirce's criterion" given on pages 13, 24, and 29, two more may be added, namely to: Mansfield Merriman's "List of writings, related to the method of least squares, with historical and critical notes," *Transactions of the Connecticut Academy*, vol. 4, 1877, especially pages 192–222; and to J. W. L. Glaisher's "On the law of facility of errors of observations and on the method of least squares," *Memoirs of the Royal Astronomical Society*, vol. 39, 1872, especially pages 120–121; compare with this page 397 of Glaisher's paper "On the rejection of discordant observations," *Mo. Notices R. Astr. Soc.*, vol. 33. The present Monograph seems to be the only one which refers to Peirce's 1878 paper on his criterion (which is followed by C. A. Schott's statement as to its value).

Gould's paper of 1855 was first published as follows: "Report to Professor A. D. Bache" . . . "containing directions and tables for the use of Peirce's criterion for the rejection of doubtful observations," *Coast Survey, Report for 1854*, Washington, 1855, pp. 131*–138*.

BENJAMIN PEIRCE'S
LINEAR ASSOCIATIVE ALGEBRA
AND C.S. PEIRCE

Raymond Clare Archibald

BENJAMIN PEIRCE'S LINEAR ASSOCIATIVE ALGEBRA AND C. S. PEIRCE

By RAYMOND CLARE ARCHIBALD, Brown University

Since the greater part of an issue of this Monthly (January, 1925) was devoted to the life and work of Benjamin Peirce, it would seem appropriate to place on record in the same publication a vigorous document of his very able son, the late C. S. Peirce. This document, dated June 28, 1910, is a two page sheet which was in his copy of Jordan's *Traité des Substitutions et des Equations Algébriques*. It was written by Peirce in his seventy-first year. Except for the footnotes, which I have added, the transcription of the document is as follows:

"I will record a reminiscence about this book. It was published in 1870, the same year as the date of the original edition of my father's *Linear Associative*

Algebra (though I am sure this was not lithographed for a year or two after the general theory was complete). I had first put my father up to that investigation by persistent hammering upon the desirability of it. There was one feature of this work, however, which I never could approve of, and in vain endeavoured to get him to change. It was his making his coefficients, or scalars, to be susceptible of taking imaginary values. In vain I represented to him that the system of imaginary quantity has two dimensions, and is consequently a double algebra. But it was always next to impossible to induce him to take a logical view of any subject. He did me the honor to reply to my arguments in a footnote on p. 19 of the Ed. of 1870 (p. 9 of that of 1882).[1] The reply is pure bosh. His "broad philosophy" which could not be definitely expressed, was a mere habit of feeling. He was a creature of feeling, and had a superstitious reverence for "the square root of minus one"; and as to the absence of it "trammeling" research, that only means that he was not in possession of any machinery for dealing with the problems that lie- beyond its scope. If Hamilton had done as he would have had him, the calculus of quaternions could not have come into being, because division would not generally have had a determinate result. The substance of this work of Jordan was inaccessible to mathematicians who did not choose to devote near a life-time to it, before the work appeared, but if my father had been able to acquaint himself with the Galois theory of equations, and had taken advantage of the possibility, he certainly must have come to see that I had been quite right in my contention. I happened to be in London in 1870 and coming across the book on Hachette's counter purchased it. But I turned it over to the Coast Survey and so was subsequently forced to surrender it along with much else that maimed me intellectually. The dirty fellows who played me this trick got nothing by it except the pleasure of harming me. My activities not lying in the direction of mathematics, I never, while I had the book, got time to master it. When I found that my brother[2] had purchased the book in 1874, I told him it was the very book he needed to study, and that he would get a flood of illumination

---

[1] C. S. Peirce was the editor, with "addenda and notes," of the second edition of his father's *Linear Associative Algebra*. The footnote here referred to is as follows: "Hamilton's total exclusion of the imaginary of ordinary algebra from the calculus as well as from the interpretation of quaternions will not probably be accepted in the future development of this algebra. It evinces the resources of his genius that he was able to accomplish his investigations under these trammels. But like the restrictions of the ancient geometry, they are inconsistent with the generalizations and broad philosophy of modern science. With the restoration of the ordinary imaginary, quaternions becomes Hamilton's biquaternions. From this point of view, all the algebras of this research would be called bi-algebras. But with the ordinary imaginary is involved a vast power of research, and the distinction of names should correspond; and the algebra which loses it should have its restricted nature indicated by such a name as that of *semi-algebra*."

[2] J. M. Peirce, for forty-five years professor of mathematics at Harvard University.

from it. But he only cut the leaves of the first sheet, and remained to his dying day a superstitious worshipper of two hostile gods, Hamilton and the scalar $\sqrt{(-1)}$. As a professor of mathematics, one would have thought he might have fancied getting some insight to the mathematical advances of his day, most of which have involved the influence of this work; but that wasn't his nature. He was also largely a creature of feeling though his feelings were not of the violent kind. When he died, he left me, as his sole but sufficient legacy (I being the only poor member of the family) his mathematical books, having previously dispossessed himself of every one that he knew I particularly desired. He thought I had a copy of this."

July, 1927.

A HITHERTO UNPUBLISHED LETTER
BY
BENJAMIN PEIRCE

Jekuthiel Ginsburg

A HITHERTO UNPUBLISHED LETTER BY BENJAMIN PEIRCE

A hitherto unpublished letter now in the New York Public Library affords
an interesting glimpse of the life of Benjamin Peirce at the moment of leave-
taking from the beloved child of his brain, *Linear Associative Algebra*. Peirce's

belief in the epoch-making qualities of his creation and his anxiety over the reception which the world would give to what he considered his most important work, are evident in every line. The reader gets a vivid impression of the seer scanning the future for evidence of the nature of the response to his message.

We know now that the work was at first almost completely ignored by the world of science. In this respect, it shared the fate of a number of other genuine contributions by American scholars in the nineteenth century which failed to receive immediately the attention they deserved. Such, for example, was the case with Gibbs's work on vector analysis. The failure was often due to general distrust shown by European scholars of the achievements of American writers. This attitude, developed when there was more reason for it, was still strong at the time of the completion of Peirce's work. Another reason for failure on the part of contemporary European scholars to recognize this evidence of Peirce's originality was the pitifully inadequate means available to him for making the work known. There were only one hundred lithographed copies printed for distribution among Peirce's friends, about the poorest possible method of announcing such a message to the world. We should not, therefore, be surprised that with the exception of the English mathematician Spottiswoode, who expressed generous appreciation of Peirce's work, there was no response to Peirce's message during his lifetime. Even when the reprint of his *Linear Associative Algebra* came out in 1881 (*American Journal of Mathematics,* v. 4), it did not evoke any response on the continent. In the authoritative *Jahrbuch für die Fortschritte der Mathematik,* v. 13, the work is listed (p. 82) by title only, with a note promising that the report on the work would be given in the next volume. This promise was never kept. Thus it came about that when Scheffers and Study came out with their work, there was nobody to call the attention of scholars to the similarity of their work to that of Peirce. It was not until 1902 that an American scholar, Professor H. E. Hawkes, took upon himself to claim for Peirce the credit which was his due. Professor Hawkes's papers on the subject appeared in the *American Journal of Mathematics* (v. 24, p. 87f.), and in the *Transactions* (v. 3) of the American Mathematical Society. In the first paper, entitled "Estimate of Peirce's associative algebra", Hawkes established Peirce's priority in the matter, while in the second, entitled "On hypercomplex number systems", he enters for Peirce a claim of superiority. According to Hawkes, Peirce's ideas could be used in the development of a more powerful instrument of research than that worked out by the German scholars. This makes the failure of the world to accord Peirce an earlier recognition the more regrettable. It is noteworthy that the reviewer of Hawkes's paper in the *Jahrbuch* for 1902 seems to have overlooked this estimate of Peirce's work. In connection with this see the remarks in Professor R. C. Archibald's *Benjamin Peirce 1809-1880,* p. 16.

Of the lithographed copies distributed by Peirce in 1870, number 53 is now in the New York Public Library. This was the copy on which Peirce had based his fondest hopes, for it was one of the two copies which he sent to George Bancroft,[1] then (1870) American Ambassador to Germany, with the request that he present the other one to the Prussian Academy of Sciences. In the letter Peirce expressed his hope that the Academy would appoint a committee of geometers to pass on the merits of his work.

[1]This is the Bancroft, noted as a historian, with whom Peirce was associated as a teacher at the famous Round Hill School, Northampton, Mass., 1825-27.—EDITOR.

The following is the letter, a part of which is here reproduced in facsimile.

$18\frac{11}{8}70$

To His excellence
    Hon. George Bancroft, LL.D., Ph. D.
        Member of the Royal Academy etc. etc. etc.
My Dear Sir.

I have the honor of sending you a copy of my last work. Humble though it be, it is that upon which my future reputation must chiefly rest. It is lithographed from a manuscript and only one hundred copies are issued. A few copies thus have been distributed and each one is numbered upon the 153rd page to which I have appended my signature.

I also send you a copy for the Academy of Berlin—which I hope that you will do me the honor to present. If it would be referred to a committee of geometers for report, I should be greatly gratified. I enclose [for] you an account of some of its claims and an indication of future researches upon the same subject.

Yours Very respectfully and sincerely
Benjamin Peirce

P.S. My definition of mathematics upon the 2nd page may be very worthy of your consideration. I observe some defects of binding or paging. B. P.

A Brief Account of the Work upon Linear Associative Algebra

This work undertakes the investigation of all possible single, double, triple, quadruple, and quintuple algebras which are subject to certain simple and almost indispensable conditions. The conditions are those well known to algebraists. The terms of *distributive* and *associative* are defined on page 21. It also contains the investigation of all sextuple algebras of a certain class—i. e., of those which contain what is called in this treatise an *idempotent* element. The term idempotent is defined near the bottom of page 16. An expression A is idempotent when, like common unity, it satisfies the equation

$$A^2 = A$$

that is, when it is its own square. An algebra which does not contain an idempotent element excludes the ordinary unity and seems, therefore, to be inapplicable to any practical enquiry.

Hamilton's quaternions appear on page 59 in a very strange form with which a very curious philosophy is connected as I shall show in some subsequent memoir. This form leads by a simple induction to a natural class of algebras, of which quaternions is the simplest, and which I shall hereafter treat under the name of quaternionoidal.

In all these algebras the common imaginary is freely admitted as I state on page 19 in a note—to which especial attention should be given. By his exclusion of the imaginary from quaternions, Hamilton seems to me [to] have trammelled himself unwisely and to have shut out those broad and simple views of polarity, which are the basis of the most admirable and profound of modern investigations in algebra.

Another class of algebras which has presented itself as worthy of attention and future research may be the so-called infinitesimaloid, from their relation to the infinitesimal calculus. They occur in every class of algebra. They are found as $[a_2]$ page 47, $[a_3]$ page 49, $[a_4]$ page 55, $[a_5]$ page 73, $[a_6]$ page 134. In

signature.

I also upon send you a copy for the Academy of Berlin — which I hope that you will do me the honor to present. If it could be referred to a committee of geometers for report, I shall be specially gratified. I enclose you an account of some of its claims and an invocation of future researches upon the same subject.

Yours very respectfully and sincerely

Benjamin Peirce

P.S. My definition of mathematics upon the first page may not be unworthy of your consideration. I observe some effects of writing in passing.

B. P.

each of these cases the i may be considered as the representative of common unity, while the j is that *nilpotent* expression (see page 16) of which the infinitesimal is an imperfect representative.

I would especially draw attention to proposition 40 on page 26 and 41 on page 27 and the table on page 30, and also to proposition 50 on page 31, and proposition 54 on page 33, and lastly proposition 59 on page 37. These propositions together give a criterion by which it can be easily determined whether an algebra is pure or simply a *mixture*. They give the key to all the research.

The only systematic attempt which I have known similar to my own is given by De Morgan in a Memoir upon triple algebra, published in the *Philosophical Transactions of Cambridge* (England).[1] But he subjected his enquiries to the unnecessary condition of being commutative. (See page 22). By an investigation which I may publish at a future time, I find that his algebras are all of them capable of decomposition so that each of them is a mixture of two independent algebras, which can be of no use in any enquiry. His misfortune was that he had no criterion for the determination of the nature of an algebra, and no system of transformation and induction. He seemed to obtain many triple algebras, whereas there is only one possible triple algebra, which is subject to his conditions and that one is infinitesimaloid.

It is proved in this treatise that there can be no algebra except the ordinary one, which has not evanescent products, i. e., products which satisfy the equation $AB = 0$. Hamilton found this stumbling-block in the way of all his efforts at forming new algebras. He resorted to the . . . . . of quaternions in order to exclude it. He seems to me, thereby, to have emasculated his algebra. The evanescence must be retained as indispensable to the abstract research. This does not interfere with its rejection in the final interpretation of the results. In this respect, it corresponds to the imaginary of ordinary algebra.

Benjamin Peirce

The letter does not seem to have been published before and it contains information which may be of value to students of the life of Peirce and of that branch of mathematics with which Peirce's name is usually connected. This seems to justify its publication.

JEKUTHIEL GINSBURG

BENJAMIN PEIRCE

Sven R. Peterson

BENJAMIN PEIRCE: MATHEMATICIAN AND PHILOSOPHER

By Sven R. Peterson

Although he is nearly unknown to us, Benjamin Peirce was considered to be the greatest American mathematician of his day, and was one of the first Americans to gain an international reputation for his work in applied mathematics. He taught at Harvard for nearly fifty years (1831–1880), influenced generations of students, and played an important part in broadening the Harvard curriculum of his day. Many of his pupils became well-known mathematicians; two in particular, Chauncey Wright and his own son Charles Sanders Peirce, were instrumental in giving American philosophy a decisive new turn.[1]

The close relationship between Charles Peirce and his father has been variously described. W. B. Gallie says in a recent book:

... his real education he owed to his father, who encouraged him with his precocious laboratory experiments, and, more important, taught him mathematics. Benjamin Peirce was primarily an applied mathematician, but the originality of his mind was perhaps best shown in his *Linear Associative Algebra*, the opening sentence of which, " Mathematics is the science which draws necessary conclusions," shows an approach far in advance of current conceptions in America, and indeed in Europe. The main lines of Peirce's intellectual development were laid down by his father's teaching.[2]

This view should be contrasted with Charles Peirce's own opinion:

... it was always next to impossible to induce him to take a logical view on any subject. His " broad philosophy " which could not be definitely expressed, was a mere habit of feeling. He was a creature of feeling, and had a superstitious reverence for " the square root of minus one ". . . .[3]

---

[1] Philip P. Wiener, in *Evolution and the Founders of Pragmatism* (Harvard University Press, 1949), has clearly demonstrated the central rôle played in the development of pragmatism by the " Metaphysical Club," that group of brilliant young intellectuals who gathered in Cambridge during the early 1870's, comprising Chauncey Wright, Charles Peirce, William James, Oliver Holmes, Jr., and several others.

[2] W. B. Gallie, *Peirce and Pragmatism* (Pelican Philosophy Series, 1952), 34.

[3] Charles Sanders Peirce, in a letter quoted by Raymond Clare Archibald, "Benjamin Peirce's Linear Associative Algebra and Charles Sanders Peirce," *American Mathematical Monthly*, 34 (1927). B. Peirce himself wrote: " The imaginary square root of algebra, from which the puzzled analyst could not escape, has become the simplest reality of Quaternions, which is the true algebra of space, and clearly elucidates some of the darkest intricacies of mechanical and physical philosophy." (*Ideality in the Physical Sciences* [Boston, 1881], 29.)

An examination of Benjamin Peirce's philosophy, however, shows it to be a far-reaching evolutionary rationalism, based on the new doctrines of material evolution which had had such an impact upon traditional theistic beliefs, and resulting finally in the ideal of a single, all-inclusive science.

I

Benjamin Peirce was born in 1809, and entered Harvard as a member of the famous class of 1829, which included such men as the elder Oliver Wendell Holmes, James Freeman Clarke, Benjamin R. Curtis, and William Henry Channing. He began teaching mathematics at Harvard in 1831, at the age of twenty-two, and continued until his death in 1880.[4]

Peirce's talent for mathematics appeared early. A glimpse of him as a student is afforded by one of his classmates:

Each class had one day a week in which to take books from the college library; and I recollect that Peirce, instead of selecting novels, poetry, history, biography, or travels, as most of us did, brought back under his arm large quarto volumes of pure mathematics.[5]

Another associate, Andrew P. Peabody, some years older than Peirce, said:

Even in our senior year we listened, not without wonder, to the reports that came up to our elevated platform of this wonderful freshman, who was going to carry off the highest mathematical honors of the University.[6]

Before he was twenty, Peirce had attracted the attention of Nathaniel Bowditch, who was translating Laplace's *Mécanique Céleste,* by pointing out an error in the proof. From then on, he assisted regularly in proof-reading the copy. There were no organized observatories in America at this time; and John Quincy Adams, in his message to

---

[4] Peirce married Sarah Hunt Mills in 1833: of their four sons, the eldest, James Mills Peirce, became a prominent mathematician at Harvard; Charles Sanders Peirce, first known for his work in mathematics and physics, has received belated recognition for his discoveries in logic and philosophy; Benjamin Mills Peirce, brilliant but undisciplined, died in early manhood; Herbert Henry Davis Peirce was a Cambridge businessman. In addition to his teaching, Peirce was consulting astronomer to the American Ephemeris and Nautical Almanac (1849–1867), and Superintendent of the United States Coast Survey (1867–1874), where he displayed considerable administrative skill. He served as librarian of the College library, helped form the Harvard Observatory by lecturing on Encke's Comet in 1843, and was a charter member of the famous Saturday Club, which included Agassiz, Emerson, Holmes, Henry James, Sr., and many other notable figures.

[5] James Freeman Clarke, *Autobiography, Diary, and Correspondence,* ed. Edward Everett Hale (1891), 34.

[6] Andrew P. Peabody, *Harvard Reminiscences* (Boston, 1888), 181.

Congress of 1825, had pleaded for the establishment of an American observatory so that the Western Hemisphere would not be dependent on Europe. The publication by Bowditch of Laplace's great work stimulated interest in astronomy, and soon afterwards a number of observatories were founded.

In his early years of teaching, Peirce wrote a series of elementary textbooks in the fields of Trigonometry, Sound, Geometry, Algebra, and Mechanics. All these texts were used in his own courses at Harvard as soon as they came out, but only the Trigonometry became widely popular, the failure of the others being attributed to Peirce's condensed style and innovations in notation. These textbooks had a lasting influence on the teaching of mathematics in America, however, and their innovations eventually became commonplace. So many people complained of the difficulty in understanding them, at first, that Peirce had the dubious honor of being investigated by the Harvard Committee for Examination in Mathematics. This committee eventually reported that " the textbooks were abstract and difficult, that few could comprehend them without much explanation, that Peirce's works were symmetrical and elegant, and could be perused with pleasure by the adult mind, but that books for young students should be more simple." [7]

In addition to his published textbooks and advanced treatises,[8] Peirce wrote on a wide range of topics, mostly astronomical or physical, all in an elegant mathematical style. Some of the problems he discussed were: the motion of two adjacent pendulums, the motion of a top, the fluidity and tides of Saturn's rings, orbits for Uranus, Neptune and the 1843 comet, a new form of binary arithmetic, systems of linear and associative algebra, occultations of the Pleiades, and Espy's theory of storms.[9] This list reveals the range of Peirce's thought, and the type of mathematical problem which was important in the mid-nineteenth century. American science was just coming of age, just beginning to do original, creative work that measured up

[7] Florian Cajori, *The Teaching and History of Mathematics in the United States* (Washington, 1890), 141. When a new book came off the press, Peirce would distribute proof sheets among the students, and accept the discovery of a misprint in place of a recitation. (George F. Hoar, " Harvard College Fifty-eight Years Ago," *Scribner's Magazine*, 28 [1900], 64.)

[8] Peirce's *A System of Analytic Mechanics* (1855), was praised as far as Germany as being the best book on its subject at the time. (Cajori, *op. cit.*, 144.) Toward the end of his life, one hundred copies of the *Linear Associative Algebra* were lithographed, at the insistence of Charles Peirce, who thought it represented his father's best work.

[9] Moses King, ed. *The Harvard Register*, III (1881), 29.

to European standards, and Peirce was one of the foremost leaders
in this growth.   One young German scholar declared in 1853 that
there were no mathematicians in America, and that Peirce was the
only astronomer.[10]   Peirce was not himself an observer; he was in-
clined to think that the mathematical reduction of observations was
much more important than merely looking through a telescope, and
indeed, the great power of mathematical analysis was strikingly
brought out by the dramatic discovery of Neptune, in which Peirce
became involved.

In 1846 Leverrier, a French mathematician, seeking to explain
certain perturbations of the planet Uranus, was led to calculate the
approximate orbit and position of a new, trans-Uranian planet.   His
prediction was quickly verified by Galle in Germany, and the discovery of the new planet was everywhere hailed as a glorious triumph
for Newtonian science.   Peirce, however, raised a dissenting voice.
He pointed out that two solutions of the initial problem were possible, consisting of two entirely distinct planetary orbits.   Leverrier
had worked out only one of these orbits, and the discovered planet
Neptune, as it turned out, actually occupied the other orbit, and
would not have been discovered at all except that by chance both
predicted locations lay at that particular time in the same direction
from the earth.   Therefore, said Peirce, Leverrier had not in the
mathematical sense discovered Neptune at all.

The ensuing controversy remained somewhat academic, since after
all the planet had been found, and the theoretical preliminaries no
longer mattered.   Peirce insisted throughout that he did not mean
to detract in the slightest from Leverrier's great and laborious calculations, and that he was convinced of the correctness of those calculations, as far as they went.   The consensus of opinion by the time
of Peirce's death, however, was that both men were wrong: Leverrier
because he had simply made an error in his calculations which resulted in a wrong orbit; Peirce because he accepted this wrong orbit
as mathematically valid, and from it derived a second solution.
Leverrier had indicated the correct direction in which to look, but
had predicted the wrong distance.   Nevertheless, the net result of
the controversy was to gain for Peirce international recognition as a
mathematician and astronomer, and to increase respect for American
science in European circles.

[10] Cajori, *op. cit.*, 140.   Peirce, though characteristically modest, concurred in
this high estimate of his genius.   He was once asked what American mathematicians
thought of a recent appointment to a professorship in mathematics.   Peirce replied
that no one had a right to express an expert opinion except himself and one former
pupil, Lucien A. Wait, later professor at Cornell.   (Moses King, *op. cit.*, 127.)

Peirce, with his friend Louis Agassiz as a powerful ally, devoted much time to strengthening and liberalizing the Harvard curriculum.[11] During the 1860's, while Charles Peirce and William James were students at the Lawrence Scientific School, George Herbert Palmer was attending Harvard College, which he described in unflattering terms:

Harvard education reached its lowest point during my college course. When I entered it, it was a small and local institution with 996 students in all its departments and thirty teachers in the college Faculty. . . . Nearly all its studies were prescribed, and these were chiefly Greek, Latin, and Mathematics. There was one course in Modern History, one in Philosophy, a half-course in Economics. . . . There were two or three courses in Natural Science, taught without laboratory work. All courses were taught from textbooks and by recitations. . . . All teaching was of a low order.[12]

Peirce was successful in having mathematics made an elective, first for the Senior year only, but eventually for the whole four years. One compelling motive for this action may have been his intense dislike of teaching any but the most gifted students. When mathematics was made an elective, the students stayed away in droves, and the mathematics department became known as small, difficult, and unpopular. Generations of unhappy students have recorded what they suffered at Peirce's hands, a combination of respect for his enthusiasm and genius with a total befuddlement as to what he was trying to say. Simon Newcomb, who later became a well-known astronomer, said:

As a teacher, he was very generally considered a failure. The general view he took was that it was useless for anyone to study mathematics without a special aptitude for them; he therefore gave inapt pupils no encouragement, and made no attempt to bring his instruction within their comprehension.[13]

George F. Hoar expressed a similar opinion:

He had little respect for pupils who had not a genius for mathematics, and

[11] President Norton called these two men a team of "political men in the University administration, who worked together for the advancement of the scientific interest." (Edward Waldo Emerson, *The Early Years of the Saturday Club* [1918], 101.) Likewise, President Felton's views of education "had been much influenced by long intimacy with his next-door neighbor, Benjamin Peirce." (Charles W. Eliot, *Harvard Memories* [Cambridge, 1923], 20.)

[12] George Herbert Palmer, *The Autobiography of a Philosopher* (Boston, 1930), 12–13. Also in *Contemporary American Philosophy*, ed. George P. Adams and William P. Montague, vol. I (New York, 1930), 20.

[13] Simon Newcomb, obituary notice in *Royal Society of Edinburgh, Proceedings*, XI (1880–1882), 742.

paid little respect to them.[14]

An editorial writer for the *Springfield Republican* said:

Few men could suggest more while saying so little, or stimulate so much
while communicating next to nothing that was tangible and comprehensible.[15]

W. E. Byerly, a student of Peirce's who received the first Ph.D. from
Harvard in 1873, had this to say:

. . . he inspired rather than taught, and one's lecture notes on his courses
were apt to be chaotic. . . . Although we rarely could follow him, we certainly sat up and took notice.[16]

In view of Peirce's power to inspire and stimulate at least the
more brilliant of his students, it is interesting to note what he himself
considered to be of chief importance in education:

Enthusiasm, which is the highest element of successful instruction, can best
be imparted nearest the fountainhead, where the springs of knowledge flow
purest, and where the waters are undiluted by the weakening influence of
text book literature.[17]

Peirce worked hard at his studies, and expected his students to do the
same. He maintained his originality of thought by trying to work
out a new problem in his own way, before turning to standard works
on the subject, and whatever question was under consideration,
striving to regard it as a particular case of some more comprehensive
theorem. The one thing he distrusted most was routine method; he
strove constantly to further fresh and original thinking, sometimes
to the point where he failed to meet his students on any common
ground. As Oliver Wendell Holmes put it:

If a question interested him, he would praise the questioner, and answer
it in a way, giving his own interpretation to the question. If he did not
like the form of the student's question, or the manner in which it was asked,

[14] George F. Hoar, "Harvard College Fifty-eight Years Ago," *Scribner's
Magazine*, 28 (1900), 64.

[15] *Benjamin Peirce, A Memorial Collection*, ed. Moses King (Cambridge, 1881),
containing obituaries from the *American Journal of Science* (Nov. 1880); *The
Springfield Republican* (Oct. 23, 1880); *Woman's Journal* (Oct. 23, 1880); *Nation*
(Oct. 14, 1880); and the *Boston Daily Advertiser* (Oct. 7, 1880).

[16] W. E. Byerly, in *Benjamin Peirce 1809–1880*, ed. R. C. Archibald (Oberlin,
Ohio, 1925), 5. This memorial, also printed in the *American Mathematical Monthly*
(1925), contains a nearly complete bibliography, a list of Peirce's writings, and
reminiscences by Charles W. Eliot, A. Lawrence Lowell, and W. E. Byerly.

[17] Benjamin Peirce, "The Intellectual Organization of Harvard University,"
The Harvard Register, I (1880), 77.

he would not answer it at all. . . .[18]

Peirce advocated more research and less teaching; the instructor should devote only two hours a day or less to formal teaching, so as to be free for new investigations. He thought the system of required courses was complicated, and prevented the students from studying under the best teachers, or learning, through the use of original memoirs, about the great investigations which were for him the life of the intellect.

The system that is adapted principally to compel attention to study is comparatively unfruitful, and fails to promote sound and original scholarship. As long as the instructions are limited to formal class teaching, the College must remain a higher school, and cannot deserve the name of University.[19]

In his later years, Peirce turned more and more to working with brilliant younger men, tutoring them and setting them a stimulating example. He had an especial fondness for finding comparative unknowns whose work had been overlooked.

The Civil War, which began when Peirce was fifty-two, left little apparent trace in his writings. He was at first a pro-slavery Democrat, and had many close friends in the South. One of his students, indicted for attempting to rescue a fugitive slave, told Peirce that if he were imprisoned, he would at least have time to read Laplace's treatise. Peirce replied ironically: " In that case, I sincerely wish you may be." [20] After the fall of Fort Sumter, however, Peirce became a strong Union supporter.[21]

[18] Oliver Wendell Holmes, Medical Essays, vol. IX, *The Works of Oliver Wendell Holmes*, Standard Library Edition (Boston, 1892), 147. Holmes also wrote a poem lauding the continued youth of the class of 1829, with the refrain " We're twenty! We're twenty! ", of which one verse refers to Peirce:

> That boy with the grave mathematical look
> Made believe he had written a wonderful book,
> And the Royal Society thought it was true!
> So they chose him right in; a good joke it was, too.

[19] Benjamin Peirce, *The Harvard Register*, I, 77.
[20] Florian Cajori, *op. cit.*, 143.
[21] Edward W. Emerson, *The Early Years of the Saturday Club* (1918), 254. Peirce's daughter remembered seeing Agassiz and her father talking over some bad news from the front, tears running down their cheeks (*ibid.*, 102). His political and social views are chiefly contained in " The National Importance of Social Science in the United States," *Documents of the American Social Science Association* (Boston, Dec. 1867), where Peirce said: " Regardless of individual as well as popular prejudice, we make truth our only aim. We search the secrets of the nation's good in the depths of experience, and our end is reached when we have

Peirce maintained throughout his life a keen interest in the arts, and in his younger days enjoyed acting in private theatricals and charades, tending to be too violent and impetuous in his acting, but always original. The Peirce and Agassiz families, friends and neighbors, would often take a carriage across the river to Boston, to see Warren at the Museum, Booth at the Boston Theatre, or perhaps to hear Fanny Kemble.[22] Peirce was always among the first to read a new poem or novel, or to attend a new opera.[23] His judgment was considered to be keen, and the following comment on the translation of poetry is perhaps typical of his literary style:

> It supposes the easy control of two languages, the full and delicate perception of the flavour of the words in both, and of the various effects of verbal combinations, as well as an ear tuned to the melodies of versification of two different peoples. The mind of the translator must consent to be formed into the mould of its original, to acquire his forms of speech, his rhetorical modulation and his musical cadence, to see with his perception and feel with his emotion.[24]

As might be expected, Peirce enjoyed chess, cards, and all sorts of intellectual puzzles. At one time he played chess every noon with Andrew Peabody, when they were teaching a class jointly, and Peabody was happy to recall an occasional victory.[25] There was an apocryphal legend among the students, however, to the effect that when Peirce played cards with his mathematical sons, no one ever actually played out a hand. Each mathematician would study his

---

ascertained the inviolable laws of human nature." Peirce held that we must guard against corrupt politics by having an enlightened electorate, and since the politicians themselves cannot be trusted to educate the people, it is up to the schools, which "should teach the children that their first duty and highest privilege is to become good citizens." Good citizens are those who are content with the station in life for which they are best fitted, and do not strive fruitlessly for power or popularity.

[22] Edward W. Emerson, *op. cit.*, 102.

[23] The extent of Peirce's reading outside the field of mathematics cannot be determined, for whatever opinions he encountered became dominated by his own thinking. He referred with respect to the opinions of Aristotle and Seneca on comets, and used Comte's three stages of science as a starting point for his own evolutionary theories. He mentioned the Bible often, but became increasingly free in his interpretation of it; he was aware of the archeological discoveries being made in Palestine, and referred to such scientists as Kepler, Galileo, Newton, Lamarck, Darwin, Cuvier, and Agassiz, but only in a general and conventional way.

[24] Unpublished note in Benjamin Peirce collection, American Academy of Arts and Sciences, Boston.

[25] Andrew P. Peabody, *op. cit.*, 185.

cards, make some calculations based on the theory of probabilities, and pay the winner.[26]

II

At the time Darwin's *Origin of Species* was published, Peirce was fifty, and though his close friend, Louis Agassiz, was a leading opponent of the Darwinian thesis, Peirce himself did not take any prominent part in the controversy, regarding it as primarily a dispute among naturalists. On the other hand, Peirce accepted wholeheartedly the larger evolutionary theories, and Laplace's Nebular Hypothesis became the basis of his systematic thinking. He held that even those, like Louis Agassiz, who rejected Darwin's hypothesis, had nevertheless been "profound believers in the laws of the succession of species . . . . The difference of doctrine is one of form rather than one of substance."[27] Peirce went on to point out that Darwin's teaching was not to be confused with Darwinism, just as Plato's teaching is not the same thing as Platonism. The disciples of Darwin would do better to go straight to the facts and find out for themselves, as Darwin did, rather than preach from his text.

The chief fault in the theory of evolution, Peirce said, is that it deifies a created power:

Is there not reason to apprehend that it is placing this very evolution upon the throne which can be occupied by no created power or any metaphysical abstraction? The force of evolution is as brute and unconscious as that of fire; there is no more royalty in it than in the log which Jupiter threw down to the frogs. In its descent it has made a frightful splash in the pool of science; but the world will recover from it, as it did from the dangerous doctrine of the earth's motion.[28]

God is immanent in the world, sustaining it from day to day, and evolution, a created power which therefore still necessitates a Creator, is simply the mode in which he appears to our eyes. Furthermore, he is a lawful God, represented to us by unchanging natural laws, and hence, argued Peirce, there could be no such thing as a miracle. Only a heathen deity who rejoiced in lawlessness would operate by

[26] Edward Emerson, *op. cit.*, 102. Another apocryphal legend was that Peirce inscribed all his books: "Who steals my Peirce steals trash." Oliver Wendell Holmes (*op. cit.*, 147) recorded the story that Peirce considered Poisson's famous *Théorie du Calcul des Probabilités* to have "a distinct Poissonish, or fishy flavor running through the whole of it."

[27] Benjamin Peirce, *Ideality in the Physical Sciences* (Boston, 1881), ed. James Mills Peirce, 135. This posthumous work is the chief source of Peirce's philosophical views.

[28] *Ibid.*, 34.

breaking the law of continuity.[29]   God proclaims himself through
the rule, not through the exception; through the silent law of
gravitation, not through the whirlwind or the earthquake, except in-
sofar as these latter phenomena are also natural occurrences.   Because God has meant his eternal plan to be carried out through the
laws of evolution. and has meant man to learn that plan by studying
the universe, such uniformity of law is the only possible expression
of God's eternal meaning.   If the universe were destitute of strict
logical connection, it would be unintelligible, and hence fail to express
God's will.

Peirce was a deeply religious man, with a strong emotional allegi-
ance to certain simple theistic tenets.   Though he paid no attention
to the debates of the various sects, he clung to the fundamental
doctrine of a personal, loving God, to whom he made frequent reference in even his most technical books and papers.   In particular, he
felt that there were certain mysteries in philosophy which were beyond the reach of science, a realm of the unknowable:

Man's speculations should be subdued from all rashness and extravagance
in the immediate presence of the Creator.   And a wise philosophy will beware lest it strengthen the arms of atheism, by venturing too boldly into
so remote and obscure a field of speculation as that of the mode of creation
which was adopted by the Divine Geometer.[30]

Peirce argued that the Bible was necessary to science, to assure us
that we were not merely bestowing an arbitrary and false structure
upon a structureless chaos of atoms:

Without the faith in the Great First Cause which is derived from the Sacred
Writings, the world could have taken but few steps in Science and would

[29] Being omnipotent, God did not need to operate in space and time, and the
universe was not to be thought of as a self-operating machine whose creation had
taken place at some fixed time. " With him there is nothing distant; all objects,
celestial and terrestrial, are in immediate proximity, and the past and the future
are forever present. Deity does not exist in time and space; but they are in
him. . . ." (Ibid., 55.)

[30] Benjamin Peirce, quoted in an obituary in the Nation (Oct. 14, 1880). Peirce
often called upon the omnipotence of God as a premise which was not to be
denied.   He held perpetual motion to be impossible, for example, because " it
would have proved destructive to human belief in the spiritual origin of force,
and the necessity of a First Cause superior to matter, and would have subjected
the grand plans of Divine benevolence to the will and caprice of man." (Ideality
in the Physical Sciences, 32.)   In the same way, Peirce proved the doctrine that all
possible things exist somewhere in the universe. He argued that actuality was
simply God's thought, and for God not to think of all possible things would be a
limitation of His power.

have been likely to have remained forever in intellectual darkness.[31]

He gradually modified his position, however, as to the literal *truth* of the Biblical cosmology. At one time, Peirce held that the Biblical version of the Creation flatly contradicted the findings of science, that neither view was capable of being essentially modified, and that the Bible was "incomparably the higher authority, and must be sustained whatever science may report to the contrary." [32] He then went on to offer his solution, in a form not uncommon at that time, of the apparent dilemma. The scientist is entirely correct in the deductions he draws from the evidence before him—but there is nothing to prevent God from having created that evidence only yesterday! God could have created the earth in the year 4004 B.C., for example, complete with fossils, rock strata, archeological ruins, glacier deposits, and cave drawings, so that scientists studying this body of evidence could validly infer that the earth has had a long evolutionary history.

Such an attempt at reconciliation reveals the important fact that Peirce, with his training in science, could not reject outright the evidence being presented on the side of science, however much he would have liked to do so. Nor could he long cling to the theory that God had deliberately deceived his children, however pious the reason for deception might be. So in 1853 Peirce advanced another theory, namely that science and the Bible have equal authority, but in different spheres:

We acknowledge . . . that science has no authority to interfere with the Scriptures and perplex the Holy Writ with forced and impossible constructions of language. This admission does not derogate from the dignity of science; and we claim that the sanctity of the Bible is equally undisturbed by the denial that it was endowed with authority over the truths of physical science.[33]

[31] Undated manuscript entitled "Remarks to Middlesex Teachers' Association," in Benjamin Peirce collection, American Academy of Arts and Sciences, Boston. Peirce's religious feeling also overflowed into his class lectures and into his books. He would break off a demonstration in higher mathematics to prove to his astonished class the existence of God: "Gentlemen, as we study the universe we see everywhere the most tremendous manifestations of force. In our own experience we know of but one source of force, namely will. How then can we help regarding the forces we see in nature as due to the will of some omnipresent, omnipotent being? Gentlemen, there must be a God!" (W. E. Byerly, cf. note 16.)

[32] Undated manuscript entitled "Address to Massachusetts Teachers' Association," in Benjamin Peirce collection.

[33] *Address of Professor Benjamin Peirce on retiring from the presidency of the American Association for the Advancement of Science* (1853), 14. Extracts from this speech are found in Edward Waldo Emerson, *The Early Years of the Saturday Club* (1918), 105.

Under such a partition of knowledge into two non-overlapping realms, Peirce now denounced as " a monstrous absurdity " any conflict between science and religion:

How can there be a more faithless species of infidelity, than to believe that the Deity has written his word upon the material universe and a contradiction of it in the Gospel? [34]

Finally, in the last year of his life, Peirce undertook a detailed defense of the first chapter of Genesis, not as literal truth, but as a sort of poetic myth or pre-scientific account of the essential steps of the Creation, leaving man to fill in the details. Peirce listed the various sources of power or energy, declaring each in turn to be a created agent of God: 1) physical force, under the name of light; 2) the heavenly firmament, home of the regulating spiritual forces; 3) the infinite below, including the earth and its " exhaustless power of development "; 4) the infinite above, containing sun, moon, and stars, hence the source of illumination and time; 5) all created plants, animals, and man; 6) the unity, design, and harmony to be found throughout the universe. " It assures us that facts and laws are born of God; that in all fact there is law, and that the law is ascertained by the study of the fact." As to the brevity of the Mosaic creation and its apparent contradiction of the long evolutionary process, Peirce said only that God did not create in time, that his power to create must not be limited by restriction to a temporal order. Peirce's final view was that the first chapter of Genesis " may not be the revelation of an actual past, but it teaches where that revelation is to be found . . .," [35] a position considerably removed from his early insistence that the Bible was " incomparably the higher authority."

In setting forth the evolutionary pattern of the material universe, Peirce maintained that the laws of death and decay could not be extended to the spiritual realm. His main argument was that spiritual individuals, unlike physical individuals, could not be interchanged:

In the material world one atom can replace another without prejudice to the system. Tree can ever be substituted for tree, and beast for beast, each in its kind; so that species is everything, and individual nothing. . . . The spiritual individual is everything. The destruction of any soul would be an irreparable loss; nothing can be conceived more utterly at variance with the harmony of creation. It is an absolute impossibility. . . .[36]

[34] *Ibid.*, 11.
[35] *Ideality in the Physical Sciences*, 50.
[36] *Ibid.*, 187.

For the same reason, Peirce rejected the mystical ideal of spiritual release from the meshes of particular existence. The last thing he wanted was union with God, if that would mean loss of individuality; and therefore matter was an integral part of each personality, rather than a hindrance:

Were the communion between soul and soul direct and immediate, there would be no protection for thought; each man would take full possession of the thoughts of every other man, and there would be no such thing as personality and individuality. The body is needed to hold souls apart and to preserve their independence, as well as for conversation and mutual sympathy. Hence body and matter are essential to man's true existence. Without them, he must, as is supposed in the Chinese theology, be instantly absorbed into the infinite spirit. In this case, creation would be a false and unmeaning tragedy.[37]

A material body is as necessary in the next world as in this. Experience indicates that there are enough individuated souls here on earth to meet the demands of this life, but presumably in the next life there will be many more individuals required. Where will they come from? Mostly, Peirce thought, from other planets and stellar systems, where, for all we know, there are beings in a state of development far beyond our own, in whose more abundant life we may be permitted to share. Since our bodies are admirably fitted to our souls, a good man would take on a bigger and healthier body in the next life, and men like Shakespeare could be recognized at sight by their stature and nobility. Sin was not for Peirce a very serious offense, merely a violation of some material or spiritual law, and punishable by a temporary deformity of the new body, until the soul truly repents.

Thus the immutable law of evolution would lead us steadily upward, through life after life, in which we would acquire new sensory organs, new ways of experiencing things, and new possibilities of discovery and research. We could hope to transcend such minor limitations as the flickering, unstable atmosphere through which all astronomers must now make their observations, and perhaps take a ride on Halley's comet, which traverses the orbits of all the planets, and thus go on a grand tour of the solar system.[38] These speculations are important chiefly because Peirce held them to be a logical extension of the doctrine of evolution. If there is a grand, ordered sweep from the primal nebula up to today's complex yet intelligible

[37] Ibid., 184.
[38] Ibid., 187ff.

world, and if spiritual individuals are never destroyed, then Peirce could well look forward to the continuation of evolutionary processes in future lives, a panorama of infinite scientific research and achievement in which " the ultimate limits to which future perception and education may advance is possibly a mystery transcending the powers of research even of the archangels." [39]

Peirce held to a clear dualism between mind and matter, asserting that conscious life is different in kind from unconscious life, and could not possibly evolve from it. Though he did not deal in any specific way with the problem of interaction between mind and matter, Peirce rejected both idealism and materialism as alternative solutions. The mind of man could not have created the whole universe, he said, and then have shrunk to its present limited capacity. Likewise, for the conscious to have evolved out of the unconscious was to him inconceivable, for " it seems to be a gross violation of the principle of the necessity of an adequate cause for the production of an effect." [40] This basic dualism, which involved a parallel evolutionary process in each realm, was supported by two principles: 1) the work done by the body is exactly equivalent to the material used up in metabolism, proving that the mind does no work in the scientific sense, and exerts no physical influence on the body; 2) ". . . when a man wills, his brain is heated, and his arms and feet obey the intention of his will; they have no innate power of resistance. This is the law of harmonious action. It is sufficient to itself and to all the demands of healthy work and inquiry." [41]

These two principles of power and harmony were considered by Peirce to be " pre-established," not to be regarded by even the boldest evolutionist as mere results of the process of evolution itself. This Leibnizian parallelism was apparently meant to allow scope to free will, by freeing man from direct contact with the determined, mechanical universe of physics and astronomy. On the other hand, Peirce often asserted that the intellect was bound by laws too, in the sense that perfect mathematical knowledge meant perfect harmony with the determined physical universe, and that the only freedom possible was the freedom to make errors or to fall short of the ideal of correct thinking:

Man is forced to the discovery of regions of the universe whence the arbitrary is banished, and where there remains no evidence of choice, or consciousness

[39] *Ibid.*, 192.
[40] *Ibid.*, 31.
[41] *Ibid.*, 74.

associated with power. . . . Intellect has its laws, which are as undeviating as those of the physical world. Man's conceit, stupidity, and obstinacy cannot resist them. . . .[42]

Peirce never felt any contradiction between the assertion of free will on the one hand, and a universe controlled by destiny after a strict mathematical model on the other hand. For him, this particular mathematical universe did not involve blind fatalism—it was friendly to man, and by virtue of its pre-established harmony took care of the wants of man. Its rigid, predictable unfolding fulfilled the unchanging will of God and proved his wisdom. Peirce's favorite example was the great beds of coal, laid down eons before the coming of man, certainly not exerting the slightest influence upon his coming or development, and yet obviously laid down all the time just for man's use. An inanimate universe could not do this unaided, said Peirce, hence the God whose will is thus expressed must be a God of love and mercy.

Man's particular relation to the universe is that of observer of the determined, evolutionary unfolding of its history. Peirce held that the universe was completely knowable, in the sense that man could, given enough time, find out each law and process. Questions about the origin of the laws themselves, of course, rested in the inscrutable will of God.[43]

In an intelligible world, an intellect! What is either without the other? Take away the world and what is to become of the owner of the intellect? . . . And if it were not for the intellect, the intelligible world would be a vain show, which no believing heart would attribute to the author of all good. . . . Intellect is only a possession, an attribute, it is less than faith, less than love; but it belongs to the same soul, which having higher capacities of faith and love, was saved by infinite love. The soul of love knows the soul of man and his wants, and cannot fail to supply the intellectual food which he needs.[44]

Peirce thus placed a certain emphasis on the intellect as the instrument of problem-solving and of spiritual growth. In one of his best passages he pictured the evolutionary growth of knowledge from subjective relativism to objective truth:

[42] Ibid., 42. "A man can tell a lie, and thereby demonstrate his feebleness; the pagan god can deceive, because he is weak and limited. But our God, for the reason of his infinite strength and of his eternal omniscience, must forever coincide with himself. Otherwise, instead of being an infinite and necessary reality, he would be an infinite impossibility. At one and the same instant, he would be and not be God." (Ibid., 184.)

[43] Unpublished notes, Benjamin Peirce collection.

[44] Ideality in the Physical Sciences, 38.

Each observer starts from his own peculiar position, which may be far removed from others. In the dim and uncertain light, he pursues crude theories, imbued with the minimum of fact and the maximum of fancy. He is easily diverted from his course by some delusive *ignis fatuus* or some glittering generality. When the causes are obscure and the visible agents fail, he constructs fairies and genii, demons and gods, to work out the mysteries which he perceived, but cannot understand.[45]

Peirce gave various practical examples of this relativism of knowledge. Early geologists who lived on level plains would naturally favor the sedimentary theory of rock formation, while geologists who lived in mountainous country, perhaps with volcanos around, would tend to favor the igneous theory. Today we see that both these theories were but partial truths, and the new and broader theory uses impartially the facts amassed by both sides in their historic controversy. In Egypt, the formless, life-giving inundations of the Nile would naturally suggest to the early cosmologists the concept of Chaos as the mother of all things; whereas in the far northern countries, the hard struggle for existence and the destruction of rocks by frost and sun would suggest a battle of the gods. Peirce found a common pattern in nearly all ancient cosmologies—that of a primeval egg hatching and growing into a full-fledged universe. The elements of these early views were what Peirce called Chaos and Ideality, standing silent and immovable like the sphinx by the pyramid, and the cosmos resulted when to these two, imperturbable meditation and inactive mass, was added the third necessary element, Motion. These steps of the primitive cosmologies corresponded with the steps of actual evolution, not because the ancients had any special scientific knowledge, but because they had formed the conclusions of " a sound philosophy, which is ever consistent with itself and with undying truth." [46]

Another factor that slowed up the progress of problem-solving man, was the subjective element inherent in all our contact with the world, the channeling of our knowledge through the senses:

A fact is not a sound, it is not a star. It is sound heard by the ear; it is a star seen by the eye. In the simplest case, it is the spiritual recognition of material existence. One moiety of it may be of the earth, earthy; but the other moiety is essentially mental and dependent upon the constitution of mind.[47]

By this Peirce meant that man reads something of himself into

[45] *Ibid.*, 43ff.
[46] *Ibid.*, 12.
[47] *Ibid.*, 25.

nature, as when he sees pictures in cloud shapes, or faces in a glow-
ing fire. The beginnings of history are full of many such subjective
fancies, leading to superstition, myth, and wild guesses about the
nature of things. The ancients looked at the stars overhead, and
formed them into bears, dogs, hunters, and heroes.

Peirce escaped both the relativity of the observer and the sub-
jectivity of sense-knowledge by appealing to the power of the human
mind, and to the belief that the universe was single, unified, and
capable of reduction to an overall theory. He pointed out that a
dog, with senses keener than our own, takes no part in man's spiritual
or intellectual life; whereas a person like Laura Bridgman, with very
limited sense-experience, lived in a spiritual universe as full of
thought and knowledge as our own. The power of reason was such
that man could attain to the knowledge of facts not yet perceived,
and even of facts which could not be perceived. Such things as the
mathematical proof of the existence of the not-yet-observed planet
Neptune, or of light waves in the emptiness of outer space, were facts
for Peirce, " pictures on the imagination."

Peirce's chief faith was in the ultimate agreement of all observers
as to the laws of nature, however much they may differ when they
first began to speculate. The ancients observed the Pleiades, for
example, and composed pretty myths about the origin of the star
cluster. Modern science has determined that the stars in question
form a family related by a common stellar movement. This discovery Peirce held to be an objective fact, with nothing of human
fancy in it, which would be true even if no humans were alive to
observe the stars. Thus in the very stronghold of astrology, where
human subjective interpretation was once at a maximum, there now
appears indubitable evidence of intellectual order.

The conclusion in every department of science is essentially the same.
Whatever may have been the play of fancy, or the delusion of superstition,
or the allurement of profit, at the outset, the end has ever been a congre-
gation of facts, organized under law, and disciplined by geometry.[48]

[48] The fact that we could not know God directly was no difficulty for Peirce:
" The divine image, photographed upon the soul of man from the center of light,
is everywhere reflected from the works of creation. . . . How could it be otherwise?
Is it not a vagary of philosophy which erects one part of creation, and that the
least, into the authorship of the whole?—which ignores the Deity, because he is
materially invisible except in his works, wherein is his only possible mode of
manifestation? We might better assert that the star, which is only known by its
light, consists of mere rays. . . ." (Quoted in Merle Curti, *The Growth of American
Thought* [New York, 1943], 297.)

There was thus for Peirce an intimate relation between the structure of the mind—mathematics—and the structure of the universe—ideality. The reduction of complicated and contradictory occurrences to a single inclusive law was always ultimately possible, and led to an increase in spiritual power and comprehension. Thus, in the field of religion, the plurality of oriental and classical gods gradually gave way to the deeper truth of the single Christian God.

. . . what was the perpetual intercourse between gods and men but the perception by vivid minds of the amazing intellectuality inwrought into the unconscious material world? Remove the plurality of the deities, and the absurdities vanish; the myth is transformed into a glorious truth. . . .[49]

In this way, every branch of human knowledge, beginning in superstition, myth, and illusion, ended in rigid law, mathematical discipline, and ideal structure. Law and structure was not solely a product of the scientist's mind, on the contrary, the physical universe was impregnated with ideal structure, waiting for the perceptive observer to actualize it.

When the sculptor develops his Apollo or his Venus from the quarried marble, it is his own creation, and has his image stamped upon it; but the truth which the man of science extracts has an absolute character of its own, which no power of genius can transform, and which is neither attributable to accident nor born of human parentage.[50]

The correspondence between knowledge and reality was complete in both directions. Not only was every part of the physical universe expressible in terms of relatively simple mathematical laws, but every logically consistent mathematical system necessarily had its expression somewhere in nature. Peirce's favorite illustration was that of the Greeks studying the properties of the conic sections as pure geometry, only to have Kepler, centuries later, find those same conic sections reproduced in the various orbits of the solar system. Likewise, the study of complex numbers with all their mysterious properties proved to have extremely practical applications in the study of alternating currents. This far-reaching realistic faith was thus a vindication of all theoretical and abstract studies, on the ground that if an intellectual system were internally consistent and possible, God, whose power to think all possible things could not be denied, must already have actualized that system.

Wild as are the flights of unchained fancy, extravagant and even monstrous as are the conceptions of unbridled imagination, we have reason to believe

[49] *Ibid.*, 20.
[50] *Ibid.*, 26.

that there is no human thought, capable of physical manifestation and consistent with the stability of the material world, which cannot be found incarnated in Nature.[51]

It was in this spirit that Peirce worked out his *Linear Associative Algebra*, sketching out hundreds of possible algebraic systems, of which only two or three had yet found any practical applications.

III

The intimate relation in Peirce's philosophy between the structure of thought and the structure of the physical universe meant that the parallel developments in the realms of matter and spirit formed a single cosmology, a vast orderly cosmos infused with ideality. The simplicity of the primal chaos is gradually replaced by the simplicity of the mathematical laws which express its orderly unfolding, so that in a sense nothing changes, though there is a progressive elevation from material to spiritual simplicity, as man moves steadily closer to the God who is at the heart of the whole.

The theological stage becomes perfect when all the gods are reduced to one God; the metaphysical stage is perfect when all the abstractions are comprehended in the one abstraction of Nature; the positive stage will be perfect when all facts are resolved into one fact.[52]

Far from using Comte's formula in a positivistic way, Peirce completed the circle by identifying the ultimate fact with God himself, on the ground that no other entity could contain all the infinite complexity that must go into it.

In general, Peirce's system of evolution was based on the Nebular Theory, and his detailed treatment of its elements shows the influence of his mathematical researches:

1) *Chaos* had for Peirce a definite mathematical meaning—that set of initial conditions from which, under the differential equations of dynamics, nothing followed. It was the initial uniform distribution of matter and energy throughout all space, a sort of mist without heat, light, collision of particles, change of density, or useful energy of any sort.[53]

2) *Matter* was in itself inert, and without powers or properties except that of receiving and retaining any amount of impressed mechanical force. Matter could not originate force, or modify any force applied to it; only in this way could physical bodies be faithful

[51] *Ibid.*, 29.
[52] *Ibid.*, 11.
[53] *Ibid.*, 58.

records of the forces acting upon them.[54] In precisely the same way, the human body was inert, and incapable of resisting the human will, or else it could not be a trustworthy medium of communication between souls:

Matter must transmit thought precisely as it is received, without suspicion of transformation. It must be competent to receive all forms of thought, and incapable of resistance to any.[55]

3) *Force* was for Peirce the great wonder worker, which gave the universe all its dynamic properties, proved the existence of God, and made the world a living, friendly place. Since man can by conscious effort " exert force " and move his body, he concludes by analogy that all force has a spiritual origin. Force is in itself invisible, of course, and known only through its effects on matter.

4) *Motion* is the chief characteristic of the universe, then, because it is the sole clue to the presence of both force and matter, and because without motion there would be no events at all. The act of creation is really the creation of motion. " The earth would have remained without form, and void, if the Spirit of God had not first *moved* upon the face of the waters." [56]

Motion is the divine energy of creation; it signifies change and phenomena, and the genesis of the powers of evolution, with the controlling, planning, and warring gods. How this creative energy could itself have been born is of all mysteries the most incomprehensible. It is the inevitable impossibility inherent in any speculation which would develop everything out of nothing.[57]

5) *Equilibrium* of forces is an important element of any Newtonian system. In inorganic systems, a balance of forces produces no change in motion, whereas unstable equilibrium results in an acceleration. The striking thing about organic or living systems, Peirce

[54] A similar view of the nature of matter is found in an article by Lovering in *The Cambridge Miscellany of Mathematics, Physics, and Astronomy* (Cambridge, No. 1, April 1842, 34. This short-lived journal was edited by Peirce. Lovering wrote: " If we analyze our knowledge of matter, we shall find it to be nothing more than a record of the laws, according to which certain forces, proceeding from various centers, act upon each other. The observation and classification of these forces mark the progress of science. None of our senses ever go behind these forces, and are unable to answer the question, whether they have a substantial basis, or proceed simply from an ideal center. This is all that Berkeley, Reid, and other skilful metaphysicians could mean in denying the existence of matter."

[55] *Ideality*, 185.

[56] Benjamin Peirce, *A System of Analytic Mechanics*, 1.

[57] *Ideality*, 43.

held, was just their ability to remain in unstable equilibrium without destructive motion taking place.

Given the primal chaos with its homogeneous distribution of matter and energy, Peirce pointed out that the slightest speck of discontinuity would suffice to start nebular evolution, by providing a focus for gravitational attraction. More and more particles would be attracted to a common center, internal pressures would appear, heat would be generated, and radiant energy released. The constantly shrinking gas cloud would begin to revolve, to throw off rings of matter, and to crystallize out into suns and planets, which would then continue on their weary path of evolution.

Peirce likened this process to a pendulum free to move in a complete circle, precariously balanced at the highest point of its swing. What would now happen if an angel gave the pendulum an *infinitely small* push? For an infinitely long time, there would be no visible result, until finally, when the motion became visible, the pendulum would swing through its complete circle in a relatively short time, and appear to come to rest. In reality, however, it would be slowly approaching its point of rest, taking another infinity of time to reach it. Finally, since the pendulum still possessed its infinitesimal increment of extra energy, it would invisibly pass its equilibrium point, and the cycle would start over again, to be repeated again and again without end.

Hence, reasoned Peirce, God need only exert one infinitely small force somewhere in the primeval chaos, which would then endure without apparent change for an infinitely long time. Suddenly, in a period that would seem long to our finite calendars, but only a tick of the clock of eternity, the chaos would develop into a nebula, the nebula would condense into stars and planets, on some of which life would flash into momentary existence. In the merest blink of the cosmic eye, the suns would be dead, all energy diffused, and all creation would slowly and invisibly return to its primal state, destined to repeat the entire process, once the original impetus introduced by God had travelled its long road home.

Curiously enough, Peirce never carried his speculations to the point of considering this infinite repetition of the evolutionary process, nor did he adopt the alternative view that the cosmic unfolding would come to a halt because of the action of entropy. The action of Peirce's pendulum requires that the original energy, however small, imparted to it must be preserved not only in quantity but also in quality or form. If part of the energy goes into warming the pivot of the pendulum, or any other frictional loss, then its action will ultimately come to a halt. Peirce mentioned the subject only once:

. . . there are paths of work which cannot be retraced, and in which there is an absorption of energy which cannot be recovered. If this be so, there is energy that is practically lost; or, in other words, there is a continual diminution of the available energy of the physical world.[58]

If Peirce accepted this view at all, which seems unlikely in view of his belief that man's immortal soul requires at all times a material body to preserve his individuality, he must have believed that after all the plans of God for men would be carried out in a *finite* time:

The law of progress will not be suddenly interrupted; it cannot cease without previous indication and some evident diminution in the rate of advance. We have a just and abiding faith that our planet will endure for the development of our children and our children's children, to the latest generation.[59]

Peirce held that there was a parallel evolution in the realm of mind, the growth of a great organized system of knowledge about the universe, ultimately reducible to a single set of postulates. Peirce not only had a magnificent vision of the unification of all the sciences, but he saw clearly the nature of the deductive system which must underlie such unification. For him, a theory was simply the train of necessary consequences flowing from a given set of assumptions; hence the theory stands or falls according to its logical structure alone. Peirce's realism appears in his profound conviction that consistent postulate systems are not empty, but are found sooner or later to correspond to physical reality. In this way even the most abstract science is a revelation of God's ordered plan. Science is the highest form of worship of God, a growing and evolving worship, as nature is itself an evolving and growing expression of God.

Peirce divided science into four divisions, ranged according to a progressive increase in deductive structure:

1) *History* is the collection of all facts in a given discipline.

2) *Natural history* is the reduction of history to *necessary* facts.

3) *Natural philosophy* is the reduction of natural history to the least number of necessary facts, that is, to a set of independent postulates.

4) *Mathematics*, finally, is the set of necessary inferences from the postulates, the deductive system that follows once the postulates are given. The task of philosophy is complete when the observed

[58] *Ibid.*, 175.
[59] *Ibid.*, 181.

facts—the history—coincide throughout with the theoretical system of deductions—the mathematics.[60]

The history of astronomy is an excellent illustration of these four categories. Tycho Brahe amassed a great number of observations and empirical tables upon the motions of the planets, but had no satisfactory theory to explain those motions. Kepler reduced this "history" to a "natural history" by means of his famous three laws of motion. In another domain, Galileo reduced the history of falling objects on the earth to a natural history, by means of his laws of dynamics. Newton combined these various natural histories into a single natural philosophy, by formulating a truly universal law of gravitation. The purely mathematical deductions from this one fundamental law give us a theoretical map or image of the planetary system, and the great achievement of the nineteenth century was to prove that this theoretical map corresponded almost exactly with the real physical world. Peirce thought, along with many of his contemporaries, that the map did correspond exactly, that Newton's theoretical world *was* the world in which we live. Hence Peirce could assert with confidence that, in the field of astronomy, the main job of science was finished.

Mathematics for Peirce always meant strictly organized facts; facts without organization were nothing but curious and trivial data. "Isolated fact . . . can be committed to memory and repeated like gossip, but it overloads the intellect like green fruit in the stomach." [61] Facts combined into formulas and formulas into theory penetrate the whole domain of physical science,[62] clarifying its nature and structure, reflecting its reality.

Once established by luck and by hard work, a theory becomes a practical working device whereby to obtain new facts and new information about nature, but the theory is not itself a fact to be believed or disbelieved:

[60] Benjamin Peirce collection, unpublished manuscript.

[61] *Ibid.* In another unpublished letter, Peirce wrote: "The mathematics are an organized system of ideas, which were, however, born of facts. . . . You will perceive that my ideas of the mathematics are more comprehensive than those which are currently adopted, and that I do not restrict them to the sciences which are technically known as such. . . . I feel confident that we are sooner or later to have new kinds of mathematics and very different from those which have been hitherto developed."

[62] Peirce never held that the process of extracting theory out of a set of given facts was an automatic one, or even a simple act of intelligence. In most mathematical work, the formulation of a new theory is a genuine creative act, " . . . the intellectual result of profound thought, enduring research, and fruitful imagination."

Astronomers are frequently asked whether they believe the nebular theory. The question is logically preposterous. An hypothesis may be believed or disbelieved; but a theory is an organized system of observed phenomena, which may be accepted as good and complete, or discarded on account of its defects, but of which belief or disbelief cannot properly be predicated.[63]

The consistent goal of Peirce's philosophy is revealed in his doctrine that science simply makes actual and explicit the intelligible structure placed in the universe by God. This structure, expressible in a few differential equations, exhibits a grand cosmological scheme, unfolding from the primeval chaos, featureless and uniform, into the whole complex sequence of world events, within which man appears with his rude theories and curious beliefs, but possessed of an irreplaceable, immortal soul. The evolution of the material universe is blind, involving nothing but inert atoms thrust here and there by mechanical forces, and yet, through the basic pre-established harmony worked into it by God, it performs two essential functions. First, it underlies and sustains the spiritual universe, by making every provision for the survival and comfort of man, long before he appears on the scene. Second, the nebula unfolds in an *intelligible* fashion, so that man can look at the universe and gradually learn all the details of God's will, and thereby fulfill his glorious destiny.

It was Peirce's deepest conviction that we live in a universe that is both friendly and knowable. That fact was for him the ultimate proof of God's love for man, just as the devotion and patience of scientists was the ultimate proof of man's love for God.

Union College.

[63] *Ideality*, 51. Peirce summarized the whole process of mathematical science as follows: "Observation supplies fact. Induction ascends from fact to law. Deduction, applying the pure logic of mathematics, reverses the process and descends from law to fact. The facts of observation are liable to the uncertainties and inaccuracies of the human senses; and the first inductions of law are rough approximations to the truth. The law is freed from the defects of observation and converted by the speculations of the geometer into exact form. But it has ceased to be pure induction, and has become ideal hypothesis. Deductions are made from it with syllogistic precision, and consequent facts are logically evolved without immediate reference to the actual events of Nature. If the results of computation coincide, not merely qualitatively, but quantitatively, with observation, the law is established as a reality, and is restored to the domain of induction. If, on the contrary, there is some inexplicable divergency between the computed and the observed facts, the law must be rejected. It cannot be accepted as an expression of the rigid theory embodied in Nature." (*Ideality*, 165.)

LINEAR ASSOCIATIVE ALGEBRA

Benjamin Peirce

Linear Associative Algebra.

By BENJAMIN PEIRCE, LL. D.

LATE PERKINS PROFESSOR OF ASTRONOMY AND MATHEMATICS IN HARVARD UNIVERSITY
AND SUPERINTENDENT OF THE UNITED STATES COAST SURVEY.

New Edition, with Addenda and Notes, by C. S. PEIRCE, Son of the Author.

[*Extracted from The American Journal of Mathematics.*]

NEW YORK : D. VAN NOSTRAND, PUBLISHER.
1882.

ERRATA.

Page 10, § 31.   The first formula should read
$$(A \pm B)\, C = AC \pm BC.$$

Page 30.   The third formula should read
$$k\,(i - k) = j.$$

Page 36.   Foot-note, second line of second paragraph, read
$$j = \frac{1}{2}\,(j_1 - \mathsf{J}k_1), \quad l = \frac{1}{2}\,(1 + \mathsf{J}i_1).$$

Page 40.   Last line of foot-note.   For e, read l.

Page 52.   Multiplication table of (l_5).   For $ji = i$, read $ji = j$.

Page 75.   Last line of foot-note, insert l, at beginning of line.

Page 86.   Foot-note.   Add that on substituting $k + \mathfrak{r}j$ for k, the algebra (aw_5) reduces to (ax_5); and the same substitution reduces (ay_5) to (az_5).

Page 91.   Last line of foot-note.   For i, read l.

PREFACE.

Lithographed copies of this book were distributed by Professor Peirce among his friends in 1870. The present issue consists of separate copies extracted from *The American Journal of Mathematics*, where the work has at length been published.*

The body of the text has been printed directly from the lithograph with only slight verbal changes. Appended to it will be found a reprint of a paper by Professor Peirce, dated 1875, and two brief contributions by the editor. The foot-notes contain transformations of several of the algebras, as well as what appeared necessary in order to complete the analysis in the text at a few points. A relative form is also given for each algebra; for the rule in *Addendum* II. by which such forms may be immediately written down, was unknown until the printing was approaching completion.

The original edition was prefaced by this dedication:

To My Friends.

This work has been the pleasantest mathematical effort of my life. In no other have I seemed to myself to have received so full a reward for my mental labor in the novelty and breadth of the results. I presume that to the uninitiated the formulae will appear cold and cheerless; but let it be remembered that, like other mathematical formulae, they find their origin in the divine source of all geometry. Whether I shall have the satisfaction of taking part in their exposition, or whether that will remain for some more profound expositor, will be seen in the future.

B. P.

---

* To page n of this issue corresponds page $n+96$ of Vol. IV. of *The Journal*.

LINEAR ASSOCIATIVE ALGEBRA.

1. Mathematics is the science which draws necessary conclusions.

This definition of mathematics is wider than that which is ordinarily given, and by which its range is limited to quantitative research. The ordinary definition, like those of other sciences, is objective; whereas this is subjective. Recent investigations, of which quaternions is the most noteworthy instance, make it manifest that the old definition is too restricted. The sphere of mathematics is here extended, in accordance with the derivation of its name, to all demonstrative research, so as to include all knowledge strictly capable of dogmatic teaching. Mathematics is not the discoverer of laws, for it is not induction; neither is it the framer of theories, for it is not hypothesis; but it is the judge over both, and it is the arbiter to which each must refer its claims; and neither law can rule nor theory explain without the sanction of mathematics. It deduces from a law all its consequences, and develops them into the suitable form for comparison with observation, and thereby measures the strength of the argument from observation in favor of a proposed law or of a proposed form of application of a law.

Mathematics, under this definition, belongs to every enquiry, moral as well as physical. Even the rules of logic, by which it is rigidly bound, could not be deduced without its aid. The laws of argument admit of simple statement, but they must be curiously transposed before they can be applied to the living speech and verified by observation. In its pure and simple form the syllogism cannot be directly compared with all experience, or it would not have required an

Aristotle to discover it. It must be transmuted into all the possible shapes in which reasoning loves to clothe itself. The transmutation is the mathematical process in the establishment of the law. Of some sciences, it is so large a portion that they have been quite abandoned to the mathematician,—which may not have been altogether to the advantage of philosophy. Such is the case with geometry and analytic mechanics. But in many other sciences, as in all those of mental philosophy and most of the branches of natural history, the deductions are so immediate and of such simple construction, that it is of no practical use to separate the mathematical portion and subject it to isolated discussion.

2. The branches of mathematics are as various as the sciences to which they belong, and each subject of physical enquiry has its appropriate mathematics. In every form of material manifestation, there is a corresponding form of human thought, so that the human mind is as wide in its range of thought as the physical universe in which it thinks. The two are wonderfully matched. But where there is a great diversity of physical appearance, there is often a close resemblance in the processes of deduction. It is important, therefore, to separate the intellectual work from the external form. Symbols must be adopted which may serve for the embodiment of forms of argument, without being trammeled by the conditions of external representation or special interpretation. The words of common language are usually unfit for this purpose, so that other symbols must be adopted, and mathematics treated by such symbols is called *algebra*. Algebra, then, is formal mathematics.

3. All relations are either qualitative or quantitative. Qualitative relations can be considered by themselves without regard to quantity. The algebra of such enquiries may be called logical algebra, of which a fine example is given by Boole.

Quantitative relations may also be considered by themselves without regard to quality. They belong to arithmetic, and the corresponding algebra is the common or arithmetical algebra.

In all other algebras both relations must be combined, and the algebra must conform to the character of the relations.

4. The symbols of an algebra, with the laws of combination, constitute its *language;* the methods of using the symbols in the drawing of inferences is its *art;* and their interpretation is its *scientific application.* This three-fold analysis of algebra is adopted from President Hill, of Harvard University, and is made the basis of a division into books.

BOOK I.*

THE LANGUAGE OF ALGEBRA.

5. The language of algebra has its alphabet, vocabulary, and grammar.

6. The symbols of algebra are of two kinds: one class represent its fundamental conceptions and may be called its *letters*, and the other represent the relations or modes of combination of the letters and are called *the signs*.

7. The *alphabet* of an algebra consists of its letters; the *vocabulary* defines its signs and the elementary combinations of its letters; and the *grammar* gives the rules of composition by which the letters and signs are united into a complete and consistent system.

The Alphabet.

8. Algebras may be distinguished from each other by the number of their independent fundamental conceptions, or of the letters of their alphabet. Thus an algebra which has only one letter in its alphabet is a *single* algebra; one which has two letters is a *double* algebra; one of three letters a *triple* algebra; one of four letters a *quadruple* algebra, and so on.

This artificial division of the algebras is cold and uninstructive like the artificial Linnean system of botany. But it is useful in a preliminary investigation of algebras, until a sufficient variety is obtained to afford the material for a natural classification.

Each fundamental conception may be called a *unit;* and thus each unit has its corresponding letter, and the two words, unit and letter, may often be used indiscriminately in place of each other, when it cannot cause confusion.

9. The present investigation, not usually extending beyond the sextuple algebra, limits the demand of the algebra for the most part to six letters; and the six letters, i, j, k, l, m and n, will be restricted to this use except in special cases.

10. *For any given letter another may be substituted*, provided a new letter represents a combination of the original letters of which the replaced letter is a necessary component.

For example, any combination of two letters, which is entirely dependent for its value upon both of its components, such as their sum, difference, or product, may be substituted for either of them.

---

* Only this book was ever written. [C. S. P.]

This *principle of the substitution of letters* is radically important, and is a leading element of originality in the present investigation; and without it, such an investigation would have been impossible. It enables the geometer to analyse an algebra, reduce it to its simplest and characteristic forms, and compare it with other algebras. It involves in its principle a corresponding substitution of *units* of which it is in reality the formal representative.

There is, however, no danger in working with the symbols, irrespective of the ideas attached to them, and the consideration of the change of the original conceptions may be safely reserved for the *book of interpretation.*

11. In making the substitution of letters, the original letter will be preserved with the distinction of a subscript number.

Thus, for the letter i there may successively be substituted i_1, i_2, i_3, etc. In the final forms, the subscript numbers can be omitted, and they may be omitted at any period of the investigation, when it will not produce confusion.

It will be practically found that these subscript numbers need scarcely ever be written. They pass through the mind, as a sure ideal protection from erroneous substitution, but disappear from the writing with the same facility with which those evanescent chemical compounds, which are essential to the theory of transformation, escape the eye of the observer.

12. A *pure* algebra is one in which every letter is connected by some indissoluble relation with every other letter.

13. When the letters of an algebra can be separated into two groups, which are mutually independent, it is a *mixed algebra*. It is mixed even when there are letters common to the two groups, provided those which are not common to the two groups are mutually independent. Were an algebra employed for the simultaneous discussion of distinct classes of phenomena, such as those of sound and light, and were the peculiar units of each class to have their appropriate letters, but were there no recognized dependence of the phenomena upon each other, so that the phenomena of each class might have been submitted to independent research, the one algebra would be actually a mixture of two algebras, one appropriate to sound, the other to light.

It may be farther observed that when, in such a case as this, the component algebras are identical in form, they are reduced to the case of one algebra with two diverse interpretations.

The Vocabulary.

14. Letters which are not appropriated to the alphabet of the algebra *
may be used in any convenient sense. But it is well to employ *the small letters*
for expressions of common algebra, and *the capital letters* for those of the algebra
under discussion.

There must, however, be exceptions to this notation; thus the letter D will
denote the derivative of an expression to which it is applied, and Σ the summa-
tion of cognate expressions, and other exceptions will be mentioned as they
occur. Greek letters will generally be reserved for angular and functional
notation.

15. The three symbols J, $\ominus$, and $\ominus$ will be adopted with the signification

$$\mathsf{J} = \sqrt{-1}$$

$\ominus =$ the ratio of circumference to diameter of circle $= 3.1415926536$
$\ominus =$ the base of Naperian logarithms $= 2.7182818285,$

which gives the mysterious formula

$$\mathsf{J}^{-\mathsf{J}} = \sqrt{\ominus\,^{\ominus}} = 4.810477381.$$

16. All the signs of common algebra will be adopted; but any signification
will be permitted them which is not inconsistent with their use in common
algebra; so that, if by any process an expression to which they refer is reduced
to one of common algebra, they must resume their ordinary signification.

17. The sign $=$, which is called that of equality, is used in its ordinary sense
to denote that the two expressions which it separates are the same whole,
although they represent different combinations of parts.

18. The signs $>$ and $<$ which are those of inequality, and denote "more
than" or "less than" in quantity, will be used to denote the relations of a whole
to its part, so that the symbol which denotes the part shall be at the vertex of
the angle, and that which denotes the whole at its opening. This involves the
proposition that the smaller of the quantities is included in the class expressed
by the larger. Thus

$$B < A \ \text{ or } \ A > B$$

denotes that A is a whole of which B is a part, so that all B is A.†

---

* See § 9.
† The formula in the text implies, also, that some A is not B. [C. S. P.]

If the usual algebra had originated in qualitative, instead of quantitative, investigations, the use of the symbols might easily have been reversed; for it seems that all conceptions involved in A must also be involved in B, so that B is more than A in the sense that it involves more ideas.

The combined expression

$$B > C < A$$

denotes that there are quantities expressed by C which belong to the class A and also to the class B. It implies, therefore, that some B is A and that some A is B.* The intermediate C might be omitted if this were the only proposition intended to be expressed, and we might write

$$B > < A.$$

In like manner the combined expression

$$B < C > A$$

denotes that there is a class which includes both A and B,† which proposition might be written

$$B < > A.$$

19. A vertical mark drawn through either of the preceding signs reverses its signification. Thus

$$A \neq B$$

denotes that B and A are essentially different wholes;

$$A \not> B \ \text{ or } \ B \not< A$$

denotes that all B is not A, ‡ so that if they have only quantitative relations, they must bear to each other the relation of

$$A = B \ \text{ or } \ A < B.$$

20. The sign $+$ is called *plus* in common algebra and denotes *addition.* It may be retained with the same name, and the process which it indicates may be called addition. In the simplest cases it expresses a mere mixture, in which

---

* This, of course, supposes that C does not vanish. [C. S. P.]
† The universe will be such a class unless A or B is the universe. [C. S. P.]
‡ The general interpretation is rather that either A and B are identical or·that some B is not A. [C. S. P.]

the elements preserve their mutual independence. If the elements cannot be mixed without mutual action and a consequent change of constitution, the mere union is still expressed by the sign of addition, although some other symbol is required to express the character of the mixture as a peculiar compound having properties different from its elements. It is obvious from the simplicity of the union recognized in this sign, that the order of the admixture of the elements cannot affect it; so that it may be assumed that

$$A + B = B + A$$

and

$$(A + B) + C = A + (B + C) = A + B + C.$$

21. The sign — is called *minus* in common algebra, and denotes *subtraction*. Retaining the same name, the process is to be regarded as the reverse of addition; so that if an expression is first added and then subtracted, or the reverse, it disappears from the result; or, in algebraic phrase, it is *canceled*. This gives the equations

$$A + B - B = A - B + B = A$$

and

$$B - B = 0.$$

The sign minus is called the negative sign in ordinary algebra, and any term preceded by it may be united with it, and the combination may be called a *negative term*. This use will be adopted into all the algebras, with the provision that the derivation of the word negative must not transmit its interpretation.

22. The sign × may be adopted from ordinary algebra with the name of the sign of *multiplication*, but without reference to the meaning of the process. The result of multiplication is to be called the *product*. The terms which are combined by the sign of multiplication may be called *factors;* the factor which precedes the sign being distinguished as the *multiplier*, and that which follows it being the *multiplicand*. The words multiplier, multiplicand, and product, may also be conveniently replaced by the terms adopted by Hamilton, of *facient*, *faciend*, and *factum*. Thus the equation of the product is

multiplier × multiplicand — product; *or* facient × faciend = factum.

When letters are used, the sign of multiplication can be *omitted* as in ordinary algebra.

23. When an expression used as a factor in certain combinations gives a product which vanishes, it may be called in those combinations a *nilfactor.* Where as the multiplier it produces vanishing products it is *nilfacient*, but where it is the multiplicand of such a product it is *nilfaciend.*

24. When an expression used as a factor in certain combinations over-powers the other factors and is itself the product, it may be called an *idemfactor.* When in the production of such a result it is the multiplier, it is *idemfacient*, but when it is the multiplicand it is *idemfaciend.*

25. When an expression raised to the square or any higher power vanishes, it may be called *nilpotent;* but when, raised to a square or higher power, it gives itself as the result, it may be called *idempotent.*

The defining equation of nilpotent and idempotent expressions are respectively $A^n = 0$, and $A^n = A$; but with reference to idempotent expressions, it will always be assumed that they are of the form

$$A^2 = A,$$

unless it be otherwise distinctly stated.

26. *Division* is the reverse of multiplication, by which its results are verified. It is the process for obtaining one of the factors of a given product when the other factor is given. It is important to distinguish the position of the given factor, whether it is facient or faciend. This can be readily indicated by combining the sign of multiplication, and placing it before or after the given factor just as it stands in the product. Thus when the multiplier is the given factor, the correct equation of division is

$$\text{quotient} = \frac{\text{dividend}}{\text{divisor} \times}$$

and the equation of verification is

$$\text{divisor} \times \text{quotient} = \text{dividend}.$$

But when the multiplicand is the given factor, the equation of division is

$$\text{quotient} = \frac{\text{dividend}}{\times \text{divisor}}$$

and the equation of verification is

$$\text{quotient} \times \text{divisor} = \text{dividend}.$$

27. Exponents may be introduced just as in ordinary algebra, and they may even be permitted to assume the forms of the algebra under discussion.

There seems to be no necessary restriction to giving them even a wider range and introducing into one algebra the exponents from another. Other signs will be defined when they are needed.

The definition of the fundamental operations is an essential part of the vocabulary, but as it is subject to the rules of grammar which may be adopted, it must be reserved for special investigation in the different algebras.

The Grammar.

28. Quantity enters as a form of thought into every inference. It is always implied in the syllogism. It may not, however, be the direct object of inquiry; so that there may be logical and chemical algebras into which it only enters accidentally, agreeably to § 1. But where it is recognized, it should be received in its most general form and in all its variety. The algebra is otherwise unnecessarily restricted, and cannot enjoy the benefit of the most fruitful forms of philosophical discussion. But while it is thus introduced as a part of the formal algebra, it is *subject to every degree and kind of limitation in its interpretation.*

The free introduction of quantity into an algebra does not even involve the reception of its unit as one of the independent units of the algebra. But it is probable that without such a unit, no algebra is adapted to useful investigation. It is so admitted into quaternions, and its admission seems to have misled some philosophers into the opinion that quaternions is a triple and not a quadruple algebra. This will be the more evident from the form in which quaternions first present themselves in the present investigation, and in which the unit of quantity is not distinctly recognizable without a transmutation of the form.*

29. The introduction of quantity into an algebra naturally carries with it, not only the notation of ordinary algebra, but likewise many of the rules to which it is subject. Thus, when a quantity is a factor of a product, it has the

---

* Hamilton's total exclusion of the imaginary of ordinary algebra from the calculus as well as from the interpretation of quaternions will not probably be accepted in the future development of this algebra. It evinces the resources of his genius that he was able to accomplish his investigations under these trammels. But like the restrictions of the ancient geometry, they are inconsistent with the generalizations and broad philosophy of modern science. With the restoration of the ordinary imaginary, quaternions becomes Hamilton's biquaternions. From this point of view, all the algebras of this research would be called bi-algebras. But with the ordinary imaginary is involved a vast power of research, and the distinction of names should correspond ; and the algebra which loses it should have its restricted nature indicated by such a name as that of a *semi-algebra.*

same influence whether it be facient or faciend, so that with the notation of § 14, there is the equation

$$Aa = aA,$$

and in such a product, the quantity a may be called the *coefficient.*

In like manner, terms which only differ in their coefficients, may be added by adding their coefficients; thus,

$$(a \pm b) A = aA \pm bA = Aa \pm Ab = A (a \pm b).$$

30. The exceeding simplicity of the conception of an equation involves the identity of the equations

$$A = B \text{ and } B = A$$

and the substitution of B for A in every expression, so that

$$MA \pm C = MB \pm C,$$

or that, *the members of an equation may be mutually transposed or simultaneously increased or decreased or multiplied or divided by equal expressions.*

31. How far the principle of § 16 limits the extent within which the ordinary symbols may be used, cannot easily be decided. But it suggests limitations which may be adopted during the present discussion, and leave an ample field for curious investigation.

The distributive principle of multiplication may be adopted; namely, the principle that the product of an algebraic sum of factors into or by a common factor, is equal to the corresponding algebraic sum of the individual products of the various factors into or by the common factor; and it is expressed by the equations

$$(A \pm B)C = AB \pm BC.$$
$$C(A \pm B) = CA \pm CB.$$

32. *The associative principle of multiplication* may be adopted; namely, that the product of successive multiplications is not affected by the order in which the multiplications are performed, provided there is no change in the relative position of the factors; and it is expressed by the equations

$$ABC = (AB)C = A(BC).$$

This is quite an important limitation, and the algebras which are subject to it will be called *associative.*

33. The principle that the value of a product is not affected by the relative position of the factors is called *the commutative principle*, and is expressed by the equation

$$AB = BA.$$

This principle is *not* adopted in the present investigation.

34. An algebra in which every expression is reducible to the form of an algebraic sum of terms, each of which consists of a single *letter* with a quantitative coefficient, is called *a linear algebra.*[*] Such are all the algebras of the present investigation.

35. Wherever there is a limited number of independent conceptions, a linear algebra may be adopted. For a combination which was not reducible to such an algebraic sum as those of linear algebra, would be to that extent independent of the original conceptions, and would be an independent conception additional to those which were assumed to constitute the elements of the algebra.

36. An algebra in which there can be complete interchange of its independent units, without changing the formulae of combination, is a *completely symmetrical algebra ;* and one in which there may be a partial interchange of its units is *partially symmetrical.* But the term symmetrical should not be applied, unless the interchange is more extensive than that involved in the distributive and commutative principles. An algebra in which the interchange is effected in a·certain order which returns into itself is a *cyclic algebra.*

Thus, quaternions is a cyclic algebra, because in any of its fundamental equations, such as

$$i^2 = -1$$
$$ij = -ji = k$$
$$ijk = -1$$

there can be an interchange of the letters in the order i, j, k, i, each letter being changed into that which follows it. The double algebra in which

$$i^2 = i, \quad ij = i$$
$$j^2 = j, \quad ji = j$$

---

[*] In the various algebras of De Morgan's "Triple Algebra," the distributive, associative and commutative principles were all adopted, and they were all linear. [De Morgan's algebras are "semi-algebras." See Cambridge Phil. Trans.. viii, 241.] [C. S. P.]

is cyclic because the letters are interchangeable in the order i, j, i. But neither of these algebras is commutative.

37. When an algebra can be reduced to a form in which all the letters are expressed as powers of some one of them, it may be called a *potential algebra*. If the powers are all squares, it may be called *quadratic*; if they are cubes, it may be called *cubic*; and similarly in other cases.

Linear Associative Algebra.

38. *All the expressions of an algebra are distributive, whenever the distributive principle extends to all the letters of the alphabet.*

For it is obvious that in the equation

$$(i+j)(k+l) = ik + jk + il + jl$$

each letter can be multiplied by an integer, which gives the form

$$(ai + bj)(ck + dl) = acik + bcjk + adil + bdjl,$$

in which a, b, c and d are integers. The integers can have the ratios of any four real numbers, so that by simple division they can be reduced to such real numbers. Other similar equations can also be formed by writing for a and b, a_1 and b_1, or for c and d, c_1 and d_1, or by making both these substitutions simultaneously. If then the two first of these new equations are multiplied by J and the last by -1; the sum of the four equations will be the same as that which would be obtained by substituting for a, b, c and d, $a + \mathsf{J}a_1$, $b + \mathsf{J}b_1$, $c + \mathsf{J}c_1$ and $d + \mathsf{J}d_1$. Hence a, b, c and d may be any numbers, real or imaginary, and in general whatever mixtures A, B, C and D may represent of the original units under the form of an algebraic sum of the letters i, j, k, &c., we shall have

$$(A + B)(C + D) = AC + BC + AD + BD,$$

which is the complete expression of the distributive principle.

39. *An algebra is associative whenever the associative principle extends to all the letters of its alphabet.*

For if
$$A = \Sigma(ai) = ai + a_1j + a_2k + \&c.$$
$$B = \Sigma(bi) = bi + b_1j + b_2k + \&c.$$
$$C = \Sigma(ci) = ci + c_1j + c_2k + \&c.$$

it is obvious that

$$AB = \Sigma\,(ab_1 ij)$$
$$BC = \Sigma\,(bc_1 ij)$$
$$(AB)C = \Sigma\,(ab_1 c_2 ijk) = A\,(BC) = ABC$$

which is the general expression of the associative principle.

40. *In every linear associative algebra, there is at least one idempotent or one nilpotent expression.*

Take any combination of letters at will and denote it by A. Its square is generally independent of A, and its cube may also be independent of A and A^2. But the number of powers of A that are independent of A and of each other, cannot exceed the number of letters of the alphabet; so that there must be some least power of A which is dependent upon the inferior powers. The mutual dependence of the powers of A may be expressed in the form of an equation of which the first member is an algebraic sum, such as

$$\Sigma_m(a_m A^m) = 0.$$

All the terms of this equation that involve the square and higher powers of A may be combined and expressed as BA, so that B is itself an algebraic sum of powers of A, and the equation may be written

$$BA + a_1 A = (B + a_1)A = 0.$$

It is easy to deduce from this equation successively

$$(B + a_1)\,A^m = \quad 0$$
$$(B + a_1)\,B = \quad 0$$
$$\left(-\frac{B}{a_1}\right)^2 = -\frac{B}{a_1}$$

so that $-\dfrac{B}{a_1}$ is an idempotent expression. But if a_1 vanishes, this expression becomes infinite, and instead of it we have the equation

$$B^2 = 0$$

so that B is a nilpotent expression.

41. When there is *an idempotent expression* in a linear associative algebra, it can be assumed as one of the independent units, and be represented by *one of the letters of the alphabet;* and it may be called *the basis.*

The remaining units can be so selected as to be separable into four distinct groups.

With reference to the basis, the units of the first group are idemfactors; those of the second group are idemfaciend and nilfacient; those of the third group are idemfacient and nilfaciend; and those of the fourth group are nilfactors.

First. The possibility of the selection of all the remaining units as idemfaciend or nilfaciend is easily established. For if i is the idempotent base, its definition gives

$$i^2 = i.$$

The product by the basis of another expression such as A may be represented by B, so that

$$iA = B,$$

which gives

$$iB = i^2A = iA = B$$
$$i(A - B) = iA - iB = B - B = 0,$$

whence it appears that B is idemfaciend and $A - B$ is nilfaciend. In other words, A is divided into two parts, of which one is idemfaciend and the other is nilfaciend; but either of these parts may be wanting, so as to leave A wholly idemfaciend or wholly nilfaciend.

Secondly. The still farther subdivision of these portions into idemfacient and nilfacient is easily shown to be possible by this same method, with the mere reversal of the relative position of the factors. Hence are obtained the required four groups.

The basis itself may be regarded as belonging to the first group.

42. Any algebraic sum of the letters of a group is an expression which belongs to the same group, and may be called *factorially homogeneous.*

43. *The product of two factorially homogeneous expressions, which does not vanish, is itself factorially homogeneous, and its faciend name is the same as that of its facient, while its facient name is the same as that of its faciend.*

Thus, if A and B are, each of them, factorially homogeneous, they satisfy the equations

$$i(AB) = (iA)B,$$
$$(AB)i = A(Bi),$$

which shows that the nature of the product as a faciend is the same as that of the facient A, and its nature as a facient is the same as that of the faciend B.

44. *Hence, no product which does not vanish can be commutative unless both its factors belong to the same group.*

45. *Every product vanishes, of which the facient is idemfacient while the faciend is nilfaciend ; or of which the facient is nilfacient while the faciend is idemfaciend.* For in either case this product involves the equation

$$AB = (Ai)B = A(iB) = 0.$$

46. The combination of the propositions of §§ 43 and 45 is expressed in the following form of a multiplication table. In this table, each factor is expressed by two letters, of which the first denotes its name as a faciend and the second as a facient. The two letters are d and n, of which d stands for *idem* and n for *nil.* The facient is written in the left hand column of the table and the faciend in the upper line. The character of the product, when it does not vanish, is denoted by the combination of letters, or when it must vanish, by the zero, which is written upon the same line with the facient and in a column under the faciend.

	dd	dn	nd	nn
dd	dd	dn	0	0
dn	0	0	dd	dn
nd	nd	nn	0	0
nn	0	0	nd	nn

47. It is apparent from the inspection of this table, that *every expression* which belongs to the second or third group is nilpotent.

48. It is apparent that *all commutative products which do not vanish are restricted to the first and fourth groups.*

49. It is apparent that every continuous product which does not vanish, has the same faciend name as its first facient, and the same facient name as its last faciend.

50. Since the products of the units of a group remain in the group, they cannot serve as the bond for uniting different groups, which are the necessary conditions of a pure algebra. Neither can the first and fourth groups be connected by direct multiplication, because the products vanish. *The first and fourth groups, therefore, require for their indissoluble union into a pure algebra that there should be units in each of the other two groups.*

51. In an algebra which has more than two independent units, it cannot happen that all the units except the base belong to the second or to the third group. For in this case, each unit taken with the base would constitute a double algebra, and there could be no bond of connection to prevent their separation into distinct algebras.

52. *The units of the fourth group are subject to independent discussion, as if they constituted an algebra of themselves.* There must be in this group an idempotent or a nilpotent unit. If there is an idempotent unit, it can be adopted as *the basis of this group, through which the group can be subdivided into subsidiary groups.*

The idempotent unit of the fourth group can even be made the basis of the whole algebra, and the first, second and third groups will respectively become the fourth, third and second groups for the new basis.

53. *When the first group comprises any units except the basis, there is besides the basis another idempotent expression, or there is a nilpotent expression.* By a process similar to that of § 40 and a similar argument, it may be shown that for any expression A, which belongs to the first group, there is some least power which can be expressed by means of the basis and the inferior powers in the form of an algebraic sum. This condition may be expressed by the èquation

$$\Sigma_m (a_m A^m) + bi = 0.$$

If then h is determined by the ordinary algebraic equation

$$\Sigma_m (a_m h^m) + b = 0,$$

and if

$$A_1 = A - hi$$

is substituted for A, an equation is obtained between the powers of A, from which an idempotent expression, B, or else a nilpotent expression, can be deduced precisely as in § 40.*

54. *When there is a second idempotent unit in the first group, the basis can be changed so as to free the first group from this second idempotent unit.*

Thus if i is the basis, and if j is the second idempotent unit of the first group, the basis can be changed to

---

* The equation in h may have no algebraic solution, in which case the new idempotent or nilpotent would not be a direct algebraic function of i and A.   [C. S. P.]

$$i_1 = i - j;$$

and with this new basis, j passes from the first to the fourth group. For

First, the new basis is idempotent, since

$$i_1^2 = (i - j)^2 = i^2 - 2ij + j^2 = i - j = i_1;$$

and *secondly*, the idempotent unit j passes into the fourth group, since

$$i_1 j = (i - j)j = ij - j^2 = j - j = 0,$$
$$ji_1 = j(i - j) = ji - j^2 = j - j = 0.$$

55. *With the preceding change of basis, expressions may pass from idemfacient to nilfacient, or from idemfaciend to nilfaciend, but not the reverse.*

For *first*, if A is nilfacient with reference to the original basis, it is also, by § 45, nilfacient with reference to the new basis; or if it is nilfaciend with reference to the original basis, it is nilfaciend with reference to the new basis.

Secondly, all expressions which are idemfacient with reference to the original basis, can, by the process of § 41, be separated into two portions with reference to the new basis, of which portions one is idemfacient and the other is nilfacient; so that the idemfacient portion remains idemfacient, and the remainder passes from being idemfacient to being nilfacient. The same process may be applied to the faciends with similar conclusions.

56. It is evident, then, that each group* can be reduced so as not to contain more than one idempotent unit, which will be its basis. In the groups which bear to the basis the relations of second and third groups, there are only nilpotent expressions.

57. *In a group or an algebra which has no idempotent expression, all the expressions are nilpotent.*

Take any expression of this group or algebra and denote it by A. If no power of A vanished, there must be, as shown in § 40, some equation between the powers of A of the form

$$\Sigma_m a_m A^m = 0,$$

in which a_1 must vanish, or else there would be an idempotent expression as is shown in § 40, which is contrary to the present hypothesis. If then m_0 denote

---

* That is, the first group as well as each of the subsidiary groups of § 52. [C. S. P.]

the exponent of the least power of A that entered into this equation, and $m_0 + h$ the exponent of the highest power that occurred in it, the whole number of terms of the equation would be, at most, $h + 1$. If, now, the equation were multiplied successively by A and by each of its powers as high as that of which the exponent is $(m_0 - 1)h$, this highest exponent would denote the number of new equations which would be thus obtained. If, moreover,

$$B = A^{m_0},$$

then the highest power of A introduced into these equations would be

$$A^{(m_0-1)h + m_0 + h} = A^{m_0(h+1)} = B^{h+1}.$$

The whole number of powers of A contained in the equations would be $m_0 h + 1$, and $h + 1$ of these would always be integral powers of B; and there would remain $(m_0 - 1)h$ in number which were not integral powers of B. There would be, therefore, equations enough to eliminate all the powers of A that were not integral powers of B and still leave an equation between the integral powers of B; and this would generally include the first power of B. From this equation, an idempotent expression could be obtained by the process of § 40, which is contrary to the hypothesis of the proposition.

Therefore it cannot be the case that there is any equation such as that here assumed; and therefore there can be no expression which is not nilpotent. The few cases of peculiar doubt can readily be solved as they occur; but they always must involve the possibility of an equation between fewer powers of B than those in the equation in A.[*]

58. *When an expression is nilpotent, all its powers which do not vanish are mutually independent.*

Let A be the nilpotent expression, of which the nth power is the highest which does not vanish. There cannot be any equation between these powers of the form

$$\Sigma_m a_m A^m = 0.$$

---

[*] In saying that the equation in B will *generally* include the first power of B, he intends to waive the question of whether this always happens. For, he reasons, if this is not the case then the equation in B is to be treated just as the equation in A has been treated, and such repetitions of the process must ultimately produce an equation from which either an idempotent expression could be found, or else A would be proved nilpotent. [C. S. P.]

For if m_0 were the exponent of the lowest power of A in this equation, the multiplication of the equation by the $(n-m_0)^{\text{th}}$ power of A reduces it to

$$a_{m_0} A^n = 0 , \quad a_{m_0} = 0 ,$$

that is, the $m_0{}^{\text{th}}$ power of A disappears from the equation, or there is no least power of A in the equation, or, more definitely, there is no such equation.

59. *In a group or an algebra which contains no idempotent expression, any expression may be selected as the basis; but one is preferable which has the greatest number of powers which do not vanish.* All the powers of the basis which do not vanish may be adopted as independent units and represented by the letters of the alphabet.

A nilpotent group or algebra may be said to be of the same order with the number of powers of its basis that do not vanish, provided the basis is selected by the preceding principle. Thus, if the squares of all its expressions vanish, it is of the *first order;* if the cubes all vanish and not all the squares, it is *of the second order,* and so on.

60. It is obvious that *in a nilpotent group whose order equals the number of letters which it contains, all the letters except the basis may be taken as the successive powers of the basis.*

61. In a nilpotent group, every expression, such as A, has some least power that is nilfacient with reference to any other expression, such as B, and which corresponds to what may be called *the facient order of B relatively to A ;* and in the same way, there is some least power of A which is nilfaciend with reference to B, and which corresponds to *the faciend order of B relatively to A.* When the facient and faciend orders are treated of irrespective of any especial reference, *they must be referred to the base.*

The facient order of a product which does not vanish, is not higher than that of its facient ; and the faciend order is not higher than that of its faciend.

62. After the selection of the basis of a nilpotent group, some one from among the expressions which are independent of the basis may be selected by the same method by which the basis was itself selected, *which, together with all its powers that are independent of the basis, may be adopted as new letters ;* and again, from the independent expressions which remain, *new letters may be selected by the same process, and so on until the alphabet is completed.* In making these selections, regard should be had to the factorial orders of the products.

63. In every nilpotent group, *the facient order of any letter which is indepen-dent of the basis can be assumed to be as low as the number of letters which are independent of the basis.*

Thus, if the number of letters which are independent of the basis is denoted by n', and if n is the order of the group (and for the present purpose it is suffi-cient to regard n' as being less than n), it is evident that any expression, A, with its successive products by the powers of the basis i, as high as the n'^{th}, and the powers of the basis which do not vanish, cannot all be independent of one another; so that there must be an equation of the form

$$\sum_{1}^{n}{}_{m}a_m i^m + \sum_{0}^{n'}{}_{m}b_m i^m A = 0 .$$

Accordingly, it is easy to see that there is always a value of A_1 of the form

$$A_1 = A - \sum_{1}^{n}{}_{m}h_m i^m$$

which will give

$$i^m A_1 = 0 ,$$

which corresponds to the condition of this section.

There is a similar condition which holds in every selection of a new letter by the method of the preceding section.

64. *In a nilpotent group, the order of which is less by unity than the number of letters, the letter which is independent of the basis and its powers may be so selected that its product into the basis shall be equal to the highest power of the basis which does not vanish, and that its square shall either vanish or shall also be equal to the highest power of the basis that does not vanish.* Thus, if the basis is i, and if the order of the algebra is n, and if j is the remaining letter, it is obvious, from § 63, that j might have been assumed such that

$$ij = 0 ,$$

which gives

$$iji = ij^2 = 0 ;$$

and therefore,

$$ji = ai^n + bj ,$$
$$j^2 = a'i^n + bj ,$$
$$0 = ji^{n+1} = bji^n = b^n ji = b ,$$
$$ji = ai^n ,$$
$$j^2 i = aji^n = 0 = b'j^2 = b' ,$$
$$j^2 = a'i^n ;$$

so that if
$$h = \left(\frac{a'}{a^2}\right)^{\frac{1}{n-2}}, \quad k = \left(\frac{a'^{n-1}}{a^n}\right)^{\frac{1}{n-2}}$$
$$j_1 = \frac{j}{k}, \qquad i_1 = \frac{i}{h},$$

we have
$$j_1 i_1 = i_1{}^n = j_1^2,$$

and i_1 and j_1 can be substituted for i and j, which conforms to the proposition enunciated.

It must be observed, however, that the analysis needs correction when the group is of the second order.

65. *In a nilpotent group of the first order, the sign of a product is merely reversed by changing the order of its factors.* Thus, if
$$A^2 = B^2 = (A + B)^2 = 0$$

it follows by development, that
$$(A + B)^2 = A^2 + AB + BA + B^2 = AB + BA = 0$$
$$BA = -AB,$$

which is the proposition enunciated.

66. *In general, in any nilpotent group of the nth order, if (A^s, B^t) denotes the sum of all possible products of the form*
$$A^p B^q \ A^{p'} B^{q'} \ A^{p''} B^{q''} \ \ldots$$

in which
$$\Sigma p = s, \quad \Sigma q = t,$$
and if
$$s + t = n + 1,$$
it will be found that
$$(A^s, \ B^t) = 0.$$

For since
$$(A + xB)^{n+1} = 0$$

whatever be the value of x, the multiplier of each power of x must vanish, which gives the proposed equation
$$(A^s, \ B^t) = 0.$$

67. *In the first group of an algebra, having an idempotent basis, all the expressions except the basis may be assumed to be nilpotent.* For, by the same argument as that of § 53, any equation between an expression and its successive powers and the basis must involve an equation between another expression which is

easily defined and its successive powers without including the basis. But it follows from the argument of §57, that such an equation indicates a corresponding idempotent expression; whereas it is here assumed that, in accordance with §56, each group has been brought to a form which does not contain any other idempotent expression than the basis. It must be, therefore, that all the other expressions are nilpotent.

68. *No product of expressions in the first group of an algebra having an idempotent basis, contains a term which is a multiple of the basis.*

For, assume the equation

$$AB = -xi + C,$$

in which A, B and C are nilpotents of the orders m, n and p, respectively. Then,

$$0 = A^{m+1}B = -xA^m + A^m C$$
$$A^m C = xA^m$$
$$0 = A^m C^{p+1} = xA^m C^p = x^2 A^m C^{p-1} = x^{p+1} A^m = x,$$

that is, the term $-xi$ vanishes from the product AB.

69. It follows, from the preceding section, that *if the idempotent basis were taken away from the first group of which it is the basis, the remaining letters of the first group would constitute by themselves a nilpotent algebra.*

Conversely, *any nilpotent algebra may be converted into an algebra with an idempotent basis, by the simple annexation of a letter idemfaciend and idemfacient with reference to every other.**

70. However incapable of interpretation the nilfactorial and nilpotent expressions may appear, they are obviously an essential element of the calculus of linear algebras. Unwillingness to accept them has retarded the progress of discovery and the investigation of quantitative algebras. But the idempotent basis seems to be equally essential to actual interpretation. The purely nilpotent algebra may therefore be regarded as an ideal abstraction, which requires the introduction of an idempotent basis, to give it any position in the real universe. In the subsequent investigations, therefore, the purely nilpotent algebras must be regarded as the first steps towards the discovery of algebras of a higher degree resting upon an idempotent basis.

---

* That every such algebra must be a pure one is plain, because the algebra (a_2) is so. [C. S. P.]

71. Sufficient preparation is now made for the

INVESTIGATION OF SPECIAL ALGEBRAS.

The following notation will be adopted in these researches. Conformably with §9, the letters of the alphabet will be denoted by i, j, k, l, m and n. To these letters will also be respectively assigned the numbers 1, 2, 3, 4, 5 and 6. Moreover, their coefficients in an algebraic sum will be denoted by the letters a, b, c, d, e and f. Thus, the product of any two letters will be expressed by an algebraic sum, and below each coefficient will be written in order the numbers which are appropriate to the factors. Thus,

while
$$jl = a_{24}\, i + b_{24}\, j + c_{24}\, k + d_{24}\, l + e_{24}\, m + f_{24}\, n,$$
$$lj = a_{42}\, i + b_{42}\, j + c_{42}\, k + d_{42}\, l + e_{42}\, m + f_{42}\, n.$$

In the case of a square, only one number need be written below the coefficient, thus
$$k^2 = a_3\, i + b_3\, j + c_3\, k + d_3\, l + e_3\, m + f_3\, n.$$

The investigation simply consists in the determination of the values of the coefficients, corresponding to every variety of linear algebra; and the resulting products can be arranged in a tabular form which may be called the multiplication-table of the algebra. Upon this table rests all the peculiarity of the calculus. In each of the algebras, it admits of many transformations, and much corresponding speculation. The basis will be denoted by i.

72. The distinguishing of the successive cases by the introduction of numbers will explain itself, and is an indispensable protection from omission of important steps in the discussion.

Single Algebra.

Since in a single algebra there is only one independent unit, it requires no distinguishing letter. It is also obvious that there can be no single algebra which is not associative and commutative. Single algebra has, however, two cases:

[1], when its unit is idempotent;
[2], when it is nilpotent.

[1]. The defining equation of this case is
$$i^2 = i.$$

This algebra may be called (a_1) and its multiplication table is *

[2]. The defining equation of this case is

$$i^2 = 0.$$

This algebra may be called (b_1) and its multiplication table is †

DOUBLE ALGEBRA.

There are two cases of double algebra:

> [1], when it has an idempotent expression;
> [2], when it is nilpotent.

[1]. The defining equation of this case is

$$i^2 = i.$$

By §§ 41 and 50, there are two cases:

> [1^2], when the other unit belongs to the first group;
> [12], when it is of the second group.

The hypothesis that the other unit belongs to the third group is a virtual repetition of [12].

[1^2]. The defining equations of this case are

$$ij = ji = j.$$

It follows from §§ 67 and 69, that there is a double algebra derived from (b_1) which may be called (a_2), of which the multiplication table is ‡

---

* This algebra may be represented by $i = A : A$ in the logic of relatives. See Addenda. [C. S. P.]
† This algebra takes the form $i = A : B$, in the logic of relatives. [C. S. P.]
‡ This algebra may be put in the form $i = A : A + B : B$, $j = A : B$. [C. S. P.]

$$(a_2) \quad i \quad j$$

	i	j
i	i	j
j	j	0

[12]. The defining equations of this case are, by § 41,

$$ij = j, \quad ji = 0;$$

whence, by § 46,

$$j^2 = 0.$$

A double algebra is thus formed, which may be called (b_2), of which the multiplication table is *

$$(b_2) \quad i \quad j$$

	i	j
i	i	j
j	0	0

[2]. The defining equation of this case is

$$i^n = 0,$$

in which n is the least power of i which vanishes. There are two cases:

[21], when $n = 3$;
[2^2], when $n = 2$.

[21]. The defining equation of this case is

$$i^3 = 0,$$

and by § 60,

$$i^2 = j.$$

This gives a double algebra which may be called (c_2), its multiplication table being †

---

* This algebra may be put in the form $i = A : A , j = A : B$. [C. S. P.]
† In relative form, $i = A : B + B : C, \ j = A : C$. [C. S. P.]

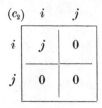

[2^2]. The defining equations of this case are

$$i^2 = j^2 = 0,$$

and it follows from §§ 64 and 65 that

$$ij = ji = 0,$$

so that there is no pure algebra in this case.*

TRIPLE ALGEBRA.

There are two cases:

[1], when there is an idempotent basis;
[2], when the basis is nilpotent.

[1]. The defining equation of this case is

$$i^2 = i.$$

There are, by §§ 41, 50 and 51, three cases:

[1^2], when j and k are both in the first group;
[12], when j is in the first, and k in the second group;
[13], when j is in the second, and k in the third group.

The case of j being in the first, and k in the third group, is a virtual repetition of [12].

[1^2]. The defining equations of this case are

$$ij = ji = j, \quad ik = ki = k.$$

---

*This case takes the form $i = A : B$, $j = C : D$. [C. S. P.]

It follows from §§ 67 and 69, that the only algebra of this case may be derived from (c_2); it may be called (a_3), and its multiplication table is *

(a_3)	i	j	k
i	i	j	k
j	j	k	0
k	k	0	0

[12]. The defining equations of this case are

$$ji = ij = j, \quad ik = k, \quad ki = 0;$$

whence, by §§ 46 and 67,

$$j^2 = k^2 = kj = 0, \quad jk = c_{23}k,$$
$$j^2k = 0 = c_{23}jk = c_{23}^2 k = c_{23} = jk,$$

and there is no pure algebra in this case.†

[13]. The defining equations of this case are

$$ij = j, \quad ki = k, \quad ji = ik = 0;$$

whence, by § 46,

$$j^2 = k^2 = kj = 0, \quad jk = a_{23} i,$$
$$jkj = 0 = a_{23}j = a_{23} = jk,$$

and there is no pure algebra in this case.‡

[2]. The defining equation of this case is

$$i^n = 0,$$

in which n is the lowest power of i that vanishes.

There are three cases:

[2₁], when $n = 4$;
[2²], when $n = 3$;
[2₃], when $n = 2$.

---

* In relative form, $i = A:A + B:B + C:C$, $j = A:B + B:C$, $k = A:C$. [U. S. P.]

† That is to say, i and j by themselves form the algebra a_2, and i and k by themselves constitute the algebra b_2, while the products of j and k vanish. Thus, the three letters are not indissolubly bound together into one algebra. In relative form, this case is, $i = A:A + B:B$, $j = A:B$, $k = A:C$. [C. S. P.]

‡ In relative form, $i = A:A + D:D$, $j = A:B$, $k = C:D$. [C. S. P.]

[21]. The defining equation of this case is

$$i^4 = 0,$$

and by § 60

$$i^2 = j, \quad i^3 = k.$$

This gives a triple algebra which may be called (b_3), the multiplication table being *

(b_3)	i	j	k
i	j	k	0
j	k	0	0
k	0	0	0

[2^2]. The defining equation of this case is

$$i^3 = 0,$$

and by §§ 59 and 64, observing the exception,

$$i^2 = j, \quad ik = 0,$$
$$ki = b_{31}j, \quad k^2 = b_3 j.$$

There is no pure algebra when b_{31} vanishes,† and there are two cases:

[$2^2 1$], when b_3 does not vanish;
[2^3], when b_3 vanishes.

[$2^2 1$]. The defining equation of this case can, without loss of generality, be reduced to

$$k^2 = j.$$

This gives a triple algebra which may be called (c_3), the multiplication table being ‡

(c_3)	i	j	k
i	j	0	0
j	0	0	0
k	aj	0	j

An interesting special example of this case is afforded by $a = -2$, when

$$i(k + i) = -j$$
$$(k + i)i = j$$
$$(k + i)^2 = 0,$$

so that $k + i$ might be substituted for k, and in this form, the multiplication table of this algebra, which may be called (c'_3), is *

(c'_3)	i	j	k
i	j	0	j
j	0	0	0
k	$-j$	0	0

---

* In relative form, $i = A:B + B:C$, $j = A:C$, $k = -A:B + B:C + A:D + D:C$.

When $a = +2$, the algebra equally takes the form (c'_3), on substituting $k - i$ for k. On the other hand, provided a is neither 2 nor -2, the algebra may be put in the form

(c''_3)	i	j	k
i	0	0	$-bj$
j	0	0	0
k	$b - \frac{1}{b}j$	0	0

To effect the transformation, we write $a = -b - \frac{1}{b}$ and substitute $i + bk$ and $i + \frac{1}{b}k$ for i and k, and $\left(b - \frac{1}{b}\right)j$ for j. Thus the algebra (c_3) has two distinct and intransmutable species, (c'_3) and (c''_3). [C. S. P.]

[2^3]. The defining equation of this case is

$$k^2 = 0,$$

and b_{31} may be reduced to unity without loss of generality, giving a triple algebra which may be called (d_3), the multiplication table being

(d_3)	i	j	k
i	j	0	0
j	0	0	0
k	j	0	0

In this case

$$(i-k)k = 0$$
$$k(i-k) = i$$
$$(i-k)^2 = 0,$$

so that $i-k$ may be substituted for i, and in this form the multiplication table is *

(d_3)	i	j	k
i	0	0	0
j	0	0	0
k	j	0	0

[23]. The defining equations of this case are

$$i^2 = j^2 = k^2 = 0,$$

and by the principles of §§ 63 and 65, it may be assumed that

$$ij = -ji = -ik = ki = 0,$$
$$jk = -kj = i.$$

---

* In relative form, $i = B : C$, $j = A : C$, $k = A : B$. This is the algebra of alio-relations in its typical form. [C. S. P.]

We thus get a triple algebra which may be called (e_3), its multiplication table being*

(e_3)	i	j	k
i	0	0	0
j	0	0	i
k	0	$-i$	0

Quadruple Algebra.

There are two cases :

        [1], when there is an idempotent basis ;
        [2], when the base is nilpotent.

[1]. The defining equation of this case is

$$i^2 = i.$$

There are six cases :

[1^2], when j, k, and l, are all in the first group ;
[12], when j and k are in the first, and l in the second group ;
[13], when j is in the first, and k and l in the second group ;
[14], when j is in the first, k in the second, and l in the third group ;
[15], when j and k are in the second, and l in the third group ;
[16], when j is in the second, k in the third, and l in the fourth group.

The other cases are excluded by §§ 50 and 51, or are obviously virtual repetitions of those which are given.

[1^2]. The defining equations of this case are

$$ij = ji = j, \quad ik = ki = k, \quad il = li = l,$$

and from §§ 60 and 69, the algebras (b_3), (c_3), (d_3), and (e_3), give quadruple algebras which may be named respectively (a_4), (b_4), (c_4), and (d_4), their multiplication tables being

---

* In relative form, $i = A : D$, $j = A : B - C : D$, $k = A : C + B : D$. This is the algebra of alternate numbers. [C. S. P.]

(a_4)	i	j	k	l
i	i	j	k	l
j	j	k	l	0
k	k	l	0	0
l	l	0	0	0

(b_4)	i	j	k	l
i	i	j	k	l
j	j	k	0	0
k	k	0	0	0
l	l	ak	0	k

(c_4)	i	j	k	l
i	i	j	k	l
j	j	k	0	0
k	k	0	0	0
l	l	k	0	0

(d_4)	i	j	k	l
i	i	j	k	l
j	j	0	0	0
k	k	0	0	j
l	l	0	$-j$	0

The special case (c'_3) gives a corresponding special case of (b_4), which may be called (b'_4), of which the multiplication table is

(b'_4)	i	j	k	l
i	i	j	k	l
j	j	k	0	k
k	k	0	0	0
l	l	$-k$	0	0

The second form of (d_3) gives a corresponding second form of (c_4), of which the multiplication table is

(c_4)	i	j	k	l
i	i	j	k	l
j	j	0	0	0
k	k	0	0	0
l	l	k	0	0

[12]. The defining equations of this case are

$$ij = ji = j, \quad ik = ki = k, \quad il = l, \quad li = 0,$$

and it follows from §§ 67 and 69, that (c_2) gives

$$j^2 = k, \quad jk = kj = k^2 = 0,$$

and from § 46,

$$lj = lk = l^2 = 0, \quad jl = d_{24}l, \quad kl = d_{34}l;$$

whence

$$j^2 l = kl = d_{24}jl = d_{24}^2 l,$$
$$jkl = d_{24}^2 jl = d_{24}^3 l = 0 = d_{24} = jl = kl,$$

and there is no pure algebra in this case.*

[13]. The defining equations of this case are

$$ij = ji = j, \quad ik = k, \quad il = l, \quad ki = li = 0,$$

which give by §§ 46 and 67

$$0 = j^2 = k^2 = kl = lk = l^2 = kj = lj;$$

and it may be assumed that

$$jk = l, \quad \text{whence} \quad jl = 0.$$

This gives a quadruple algebra which may be called (e_4), its multiplication table being †

---

* In relative form, $i = A:A + B:B + C:C + D:D$, $j = A:B + B:C$, $k = A:C$, $l = D:C$. [C. S. P.]

† In relative form, $i = A:A + B:B$, $j = A:B$, $k = B:C$, $l = A:C$. [C. S. P.]

(e_4)	i	j	k	l
i	i	j	k	l
j	j	0	l	0
k	0	0	0	0
l	0	0	0	0

[14]. The defining equations of this case are

$$ij = ji = j, \quad ik = k, \quad li = l, \quad ki = il = 0;$$

which give, by §§ 46 and 67,

$$0 = j^2 = jl = kj = k^2 = lk = l^2,$$
$$jk = c_{23}k, \quad lj = d_{42}l, \quad kl = a_{34}i + b_{34}j,$$
$$0 = j^2k = c_{23}jk = c_{23}^2k = c_{23} = jk,$$
$$0 = lj^2 = d_{42}lj = d_{42}^2l = d_{42} = lj,$$
$$0 = jkl = a_{34}j = a_{34},$$

and b_{34} cannot be permitted to vanish,* so that it does not lessen the generality to assume

$$kl = j.$$

This gives a quadruple algebra which may be called (f_4), its multiplication table being †

(f_4)	i	j	k	l
i	i	j	k	0
j	j	0	0	0
k	0	0	0	j
l	l	0	0	0

---

*For then the algebra would split up into three double algebras. [C. S. P.]

† In relative form, $i = A:A + B:B$, $j = A:B$, $k = A:C$, $l = C:B$. [C. S. P.]

[15]. The defining equations of this case are

$$ij = j, \quad ik = k, \quad li = l, \quad ji = ki = il = 0,$$

which give, by § 46,

$$0 = j^2 jk = kj = k^2 = lj = lk = l^2,$$
$$jl = a_{24}i, \quad kl = a_{34}i,$$
$$0 = jlj = a_{24}j = a_{24} = jl,$$
$$0 = kllk = a_{34}k = a_{34} = kl,$$

and there is no pure algebra in this case.*

[16]. The defining equations of this case are

$$ij = j, \quad ki = k, \quad ji = ik = il = li = 0,$$

which give, by § 46,

$$0 = j^2 = k^2 = kl = lj,$$
$$jk = a_{23}i, \quad jl = b_{24}j, \quad kj = d_{32}l, \quad lk = c_{43}k, \quad l^2 = d_4 l,$$
$$jkj = a_{23}j = d_{32}b_{24}j, \quad jlk = a_{23}b_{24}i = a_{23}c_{43}i, \quad jl^2 = b_{24}^2 j = b_{24}d_4 j,$$
$$kjk = a_{23}k = c_{43}d_{32}k, \quad kjl = d_{32}d_4 l = b_{24}d_{32}l,$$
$$lkj = d_{32}d_4 l = c_{43}d_{32}l, \quad l^2k = c_{43}^2 k = c_{43}d_4 k, \quad a_{23} = b_{24}d_{32},$$
$$0 = a_{23}(c_{43} - b_{24}) = b_{24}(b_{24} - d_{24}) = d_{32}(b_{24} - d_4) = d_{32}(c_{43} - d_4) = c_{43}(c_{43} - d_4).$$

There are two cases:

[161], when d_{32} does not vanish;
[162], when d_{32} vanishes.

[161]. The defining equation of this case can be reduced to

$$d_{32} = 1,$$

which gives

$$a_{23} = b_{24} = c_{43} = d_4.$$

There are two cases:

[161²], when d_4 does not vanish;
[1612], when d_4 vanishes.

[161²]. The defining equation of this case can be reduced to

$$d_4 = 1,$$

which gives

$$jk = i, \quad jl = j, \quad lk = k, \quad l^2 = l;$$

---

* In relative form, $i = A : A$, $j = A : B$, $k = A : C$, $l = D : A$. There are three double algebras of the form (b_2). [C. S. P.]

and there is a quadruple algebra which may be called (g_4), its multiplication table being

(g_4)	i	j	k	l
i	i	j	0	0
j	0	0	i	j
k	k	l	0	0
l	0	0	k	l

This is a form of *quaternions.**

[1612]. The defining equation of this case is

$$d_4 = 0,$$

which gives

$$jk = jl = lk = l^2 = 0,$$

---

* In relative form, $i = A : A$, $j = A : B$, $k = B : A$, $l = B : B$. This algebra exhibits the general system of relationship of individual relatives, as is shown in my paper in the ninth volume of the Memoirs of the American Academy of Arts and Sciences. In a space of four dimensions, a vector may be determined by means of its rectangular projections on two planes such that every line in the one is perpendicular to every line in the other. Call these planes the A-plane and the B-plane, and let v be any vector. Then, iv is the projection of v upon the A-plane, and lv is its projection upon the B-plane. Let each direction in the A-plane be considered as to correspond to a direction in the B-plane in such a way that the angle between two directions in the A-plane is equal to the angle between the corresponding directions in the B-plane. Then, jv is that vector in the A-plane which corresponds to the projection of v upon the B-plane, and kv is that vector in the B-plane which corresponds to the projection of v upon the A-plane.

Professor Peirce showed that we may take i_1, j_1, k_1, as three such mutually perpendicular vectors in ordinary space, that $i = \frac{1}{2}(1 - Ji_1)$, $j = \frac{1}{2}(j - Jk_1)$, $k = \frac{1}{2}(-j_1 - Jk_1)$, $l = \frac{1}{2}(1 - Ji)$. [See, also, Spottiswoode, Proceedings of the London Mathematical Society, iv, 156. Cayley, in his Memoir on the Theory of Matrices (1858), had shown how a quaternion may be represented by a dual matrix.] Thus i, j, k, l, have all zero tensors, and j and k are vectors. In the general expression of the algebra, $q = xi + yj + zk + wl$, if $x + w = 1$ and $yz = x - x^2$, we have $q^2 = q$; if $x = -w = \sqrt{-yz}$, then $q^2 = 0$. The expression $i + l$ represents scalar unity, since it is the universal idemfactor. We have, also, $Sq = \frac{1}{2}(x + w)(i + l)$, $Vq = \frac{1}{2}(x - w)i + yj + zk + \frac{1}{2}(w - x)l$, $Tq = \sqrt{\overline{xw - yz}}(i + l)$.

The resemblance of the multiplication table of this algebra to the symbolical table of §46 merits attention. [C. S. P.]

and there is a quadruple algebra which may be called (h_4), its multiplication table being *

(h_4)	i	j	k	l
i	i	j	0	0
j	0	0	0	0
k	k	l	0	0
l	0	0	0	0

[162]. The defining equation of this case is

$$d_{32} = 0,$$

which gives

$$a_{23} = 0,$$

and there can be no pure algebra for it.†

[2]. The defining equation of this case is

$$i^n = 0.$$

There are four cases:
[21], when $n = 5$;
[2²], when $n = 4$;
[23], when $n = 3$;
[24], when $n = 2$.

[21]. The defining equation of this case is

$$i^5 = 0,$$

and by § 60, $\qquad i^2 = j, \quad i^3 = k, \quad i^4 = l.$

This gives a quadruple algebra which may be called (i_4), its multiplication table being ‡

---

* In relative form, $i = A : A$, $j = A : B$, $k = C : A$, $l = C : B$. [C. S. P.]
† In this case, $i = A : A$, $l = d_4(B : B + C : C)$, $j = A : B$ or $= A : D$, $k = C : A$ or $= E : A$. [C. S. P.]
‡ In relative form, $i = A : B + B : C + C : D + D : E$, $j = A : C + B : D + C : E$, $k = A : D + B : E$, $l = A : E$. [C. S. P.]

(i_4)	i	j	k	l
i	j	k	l	0
j	k	l	0	0
k	l	0	0	0
l	0	0	0	0

[2^2]. The defining equation of this case is

$$i^4 = 0,$$

and by § 59, $$i^2 = j, \quad i^3 = k.$$

There are then, by § 64, two quadruple algebras, which may be called (j_4) and (k_4), their multiplication tables being*

(j_4)	i	j	k	l
i	j	k	0	0
j	k	0	0	0
k	0	0	0	0
l	k	0	0	k

and

(k_4)	i	j	k	l
i	j	k	0	0
j	k	0	0	0
k	0	0	0	0
l	k	0	0	0

[23]. The defining equation of this case is

$$i^3 = 0,$$

and by § 59

$$i^2 = j,$$

and it may be assumed from the principle of § 63 that

$$ik = 0,$$

which gives

$$jk = 0.$$

---

*In either of these algebras, $i = A:B + B:C + C:D$, $j = A:C + B:D$, $k = A:D$; and in (j_4) $l = A:E + E:D + A_.:C$, while in (k_4) $l = A:C$. [C. S. P.]

There are two cases: [231], when $il = k$;

[232], when $il = 0$.

[231]. The defining equation of this case is

$$il = k,$$

which gives

$jl = i^2l = ik = 0$,

$ki = a_{31}i + b_{31}j + c_{31}k + d_{31}l$,

$0 = iki = a_{31}j + d_{31}k$, $a_{31} = 0$, $d_{31} = 0$, $ki = b_{31}j + c_{31}k$.

So, because $ik^2 = 0$, $k^2 = b_3j + c_3k$,

and because $ikl = 0$, $kl = b_{34}j + c_{34}k$, $kj = kii = c_{31}ki = b_{31}c_{31}j + c_{31}^2k$.

$0 = kji = c_{31}^2 ki$, $c_{31} = 0 = kj$,

$ili = ki = b_{31}j$, $li = b_{31}i + b_{41}j + c_{41}k$, $lj = li^2 = (b_{31} + b_{31}c_{41})j$,

$0 = k^3 = c_3k^2 = c_3$, $ilk = k^2 = b_3j$, $lk = b_3i + b_{43}j + c_{43}k$,

$l^2 = a_4i + b_4j + c_4k + d_4l$, $0 = l^3 = a_4k + c_4kl + d_4l^2 = a_4li + b_4lj + c_4lk + d_4l^2$.

But kl contains no term in l, so that $d_4 = 0$.

$kl = il^2 = a_4j$, $b_{34} = a_4$, $c_{34} = 0$,

$0 = l^3 = b_{34}k + c_4b_{34}j$, $b_{34} = a_4 = 0 = kl$, $l^2 = b_4j + c_4k$,

$kil = k^2 = b_{31}jl = 0$, $0 = kli = b_{31}ki = b_{31}^2j = b_{31} = ki = lj$,

$li = b_{41}j + c_{41}k$, $lk = lil = 0$.

There are two cases:

[231²], when c_{41} does not vanish;

[2312], when c_{41} vanishes.

[231²]. The defining formula of this case is

$$c_{41} \neq 0,$$

and if p is determined by the equation

$$c_{41}p^2 + (c_4 - b_{41})p = b_4,$$

we have

$i(l + pi) = k + pj$,

$(l + pi)^2 = (c_4 + pc_{41})(k + pj)$,

so that $l + pi$ and $k + pj$ may be substituted respectively for l and k, which is the same as to make

$$b_4 = 0,$$

and there are two cases :

$$[231^3], \quad \text{when } c_4{}^* \text{ does not vanish};$$
$$[231^2 2], \quad \text{when } c_4 \text{ vanishes}.$$

$[221^3]$. The defining equation of this case can be reduced to

$$l^2 = k.$$

This gives a quadruple algebra which may be called (l_4), its multiplication table being †

(l_4)	i	j	k	l
i	j	0	0	k
j	0	0	0	0
k	0	0	0	0
l	$bj + ck$	0	0	k

$[231^2 2]$. The defining equation of this case is

$$l^2 = 0.$$

---

* *I. e.* the *new* c_4, or what has been written $c_4 + pc_{41}$. In all cases, when new letters of the alphabet of the algebra are substituted, the coefficients change with them. [C. S. P.]

†When $b = 0$, $c = 1$, we have $l(i - l) = (i - l)l = 0$; so that by the substitution of $i - l$ for i, the algebra is broken up into two of the form (c_2). When $b = 0$, $c \neq 1$, on substituting $i_1 = i - l$, $j_1 = j - ck$, $k_1 = (c - 1)^2 k$, $l_1 = (c - 1)l$, we have $i_1^2 = j_1$, $i_1 l_1 = 0$, $l_1 i_1 = l_1^2 = k_1$; so that the algebra reduces to (r_4). When $b = 1$, $c = 0$, on putting $i_1 = i - l$, $j_1 = j - k$, we have $i_1^2 = i_1 l = 0$, $l i_1 = j_1$, $l^2 = k$; so that the algebra reduces to (q_4). When $b = 1$, $c \neq 0$, on putting $i_1 = \sqrt{c^{-1}}(i - l)$, $j_1 = j + (c - 1)k$, we have $i_1^2 = l^2 = k$, $i_1 l = 0$, $l i_1 = j_1$; so that the algebra reduces to (p_4). When $b(b - 1)(bc + b - 1) \neq 0$, on putting $i_1 = (1 - b)bi - (1 - b)l$, $j_1 = (1 - b)^2(1 - b - bc)k$, $k_1 = b^2(1 - b)(1 - b - bc)j - b(1 - b)(1 - b - c + c^2 b)k$, $l_1 = b(1 - b)i - bcl$, we get the multiplication table of (o_4). When $b(b - 1) \neq 0$, $bc + b = 1$; on putting $i_1 = b(i - l)$, $j_1 = b^2(1 - b)j - b^2 ck$, $k_1 = b(1 - b - c)k$, $l_1 = bi - l$, we get the following multiplication table, which may replace that in the text:

(l_4)	i	j	k	l
i	j	0	0	j
j	0	0	0	0
k	0	0	0	0
l	k	0	0	0

In relative form, $i = A : B + B : C + A : D$, $j = A : C$, $k = A : E$, $e = A : B + D : E$. [C. S. P.]

There are two cases:

$[231^221]$, when b_{41} does not vanish;
$[231^22^2]$, when b_{41} vanishes.

$[231^221]$. The defining formula of this case is

$$b_{41} \neq 0.$$

There are two cases:

$[231^221^2]$, when $c_{41} + 1$ does not vanish;
$[231^2212]$, when $c_{41} + 1$ vanishes.

$[231^221^2]$. The defining formula of this case is

$$c_{41} + 1 \neq 0$$

so that

$$l \frac{b_{41}i + c_{41}l}{c_{41} + 1} = \frac{b_{41}^2 j + b_{41}c_{41}k}{c_{41} + 1},$$

$$\frac{b_{41}i + c_{41}l}{c_{41} + 1} l = \frac{b_{41}k}{c_{41} + 1}, \quad \left(\frac{b_{41}i + c_{41}l}{c_{41} + 1}\right)^2 = \frac{b_{41} + c_{41}}{c_{41} + 1} \cdot \frac{b_{41}j + c_{41}k}{c_{41} + 1}$$

so that the substitution of $\dfrac{b_{41}i + c_{41}l}{c_{41} + 1}$, $\dfrac{b_{41}^2 j + b_{41}c_{41}k}{c_{41} + 1}$, and $\dfrac{b_{41}k}{c_{41} + 1}$, respectively, for

i, j, and k, is the same as to assume

$$c_{41} = 0, \quad b_{41} = j,$$

which reduces this case to $[2312]$.

$[231^2212]$. The defining equation of this is easily reduced to

$$li = j - k.$$

This gives a quadruple algebra which may be called (m_4), its multiplication table being

(m_4)	i	j	k	l
i	j	0	0	k
j	0	0	0	0
k	0	0	0	0
l	$j - k$	0	0	0

PEIRCE : *Linear Associative Algebra.*

The substitution of $i - l$ and $j - k$, respectively, for i and j transforms this algebra into one of which the multiplication table is *

(m_4)	i	j	k	l
i	0	0	0	k
j	0	0	0	0
k	0	0	0	0
l	j	0	0	0

$[231^2 2^2]$. The defining equation of this case is

$$li = c_{41} k.$$

This gives a quadruple algebra which may be called (n_4), its multiplication table being †

(n_4)	i	j	k	l
i	j	0	0	k
j	0	0	0	0
k	0	0	0	0
l	ck	0	0	0

$[2312]$. The defining equation of this case is

$$li = b_{41} j,$$

which gives

$$(l - b_{41} i) i = 0,$$

so that the substitution of $l - b_{41} i$ for l passes this case virtually into $[232]$.

---

* $i = A : B + C : D$, $j = B : D$, $k = A : C$, $l = B : C$. [C. S. P.]

† In relative form, $i = A : B + B : C + D : E$, $j = A : C$, $k = A : E$, $l = B : E + c A : D$. When $c = 0$ the algebra reduces to (q_4). [C. S. P.]

[232]. The defining equation of this case is

$$il = 0,$$

and it may be assumed that

$$ki = 0,$$

$$0 = jl = kj = ik^2 = k^2i = ikl = ili = ilk = lki = il^2$$

$$k^2 = b_3 j + c_3 k + d_3 l, \quad li = b_{41} j + c_{41} k + d_{41} l,$$

$$lj = d_{41} li, \quad 0 = lji = d_{41} lj = d_{41}^2 li = d_{41} = lj.$$

There are two cases:

[2321], when c_{41} does not vanish;

[232²], when c_{41} vanishes.

[2321]. The defining equation of this case is easily reduced to

$$li = k,$$

which gives

$$0 = lik = k^2 = lil = kl$$

$$lk = l^2 i = a_4 j + d_4 k,$$

$$0 = l^3 k = d_4 l^2 k = d_4^2 lk = d_4, \quad lk = a_4 j = l^2 i,$$

$$l^2 = a_4 i + b_4 j + c_4 k,$$

$$0 = l^3 = a_4 k + c_4 a_4 j = a_4 = lk.$$

There are two cases:

[2321²], when c_4 does not vanish;

[23212], when c_4 vanishes.

[2321²]. The defining equation of this case can be reduced to

$$c_4 = 1$$

which gives a quadruple algebra which may be called (o_4), its multiplication table being *

(o_4)	i	j	k	l
i	j	0	0	0
j	0	0	0	0
k	0	0	0	0
l	k	0	0	$bj + k$

---

* In relative form, $i = A:E + E:D + B:C$, $j = A:D$, $k = A:C$, $l = A:B + B:C + bB:D$. When $b = 0$, this algebra reduces to (r_4). When $b = -1$, the substitution of $i - l$ for l reduces it to (l_4). [C. S. P.]

[23212]. The defining equation of this case is

$$l^2 = b_4 j.$$

There are two cases:

[232121], when b_4 does not vanish;

[23212²], when b_4 vanishes.

[232121]. The defining equation of this case can be reduced to

$$l^2 = j.$$

This gives a quadruple algebra which may be called (p_4), its multiplication table being *

(p_4)	i	j	k	l
i	j	0	0	0
j	0	0	0	0
k	0	0	0	0
l	k	0	0	j

[23212²]. The defining equation of this case is

$$l^2 = 0.$$

This gives a quadruple algebra which may be called (q_4), its multiplication table being †

(q_4)	i	j	k	l
i	j	0	0	0
j	0	0	0	0
k	0	0	0	0
l	k	0	0	0

---

* In relative form, $i = A:B + B:D + C:E$, $j = A:D$, $k = A:E$, $l = A:C + C:D$.   [C. S. P.]

† In relative form, $i = A:C + C:D$, $j = A:D$, $k = B:D$, $l = B:C$.   [C. S. P.]

[232²]. The defining equation of this case is

$$li = b_{41}j$$

and we have

$$k^2 = b_3 j + c_3 k + d_3 l$$
$$kl = b_{34} j + c_{34} k + d_{34} l$$
$$lk = b_{43} j + c_{43} k + d_{43} l$$
$$l^2 = b_4 j + c_4 k + d_4 l$$

so that there can be no pure algebra in this case if b_{41} vanishes,* and it may be assumed without loss of generality that

$$li = j.$$

There are two cases:

[232²1], when d_3 does not vanish;
[232³], when d_3 vanishes.

[232²1]. The defining equation of this case can be reduced to

$$k^2 = l,$$

which gives

$$0 = k^3 = kl = lk = k^2 l = l^2,$$

and there is no pure algebra in this case.†

[232³]. The defining equation of this case is

$$d_3 = 0,$$

which gives

$$0 = k^3 = c_3 k^2 = c_3, \quad k^2 = b_3 j,$$
$$0 = k^2 l = c_{34} k^2 + d_{34} kl = d_{34} = b_3 c_{34},$$
$$0 = lk^2 = c_{43} k^2 + d_{43} kl = d_{43} = b_3 c_{43}.$$

There are two cases:

[232³1], when b_3 does not vanish;
[232⁴], when b_3 vanishes.

[232³1]. The defining equation of this case can be reduced to

$$k^2 = j,$$

which gives

$$0 = c_{34} = c_{43}, \quad kl = b_{34} j, \quad lk = b_{43} j,$$
$$k(l - b_{34} j) = 0,$$

---

* In this case, j, k and l, might form any one of the algebras (b_3), (c_3), (d_3) or (e_3). [C. S. P.]
† The case is impossible because $ki = 0$ and $k^2 i = j$. [C. S. P.]

so that $l - b_{34}j$ can be substituted for l without loss of generality, which is the same as to assume

$$kl = 0 ;$$

and this gives

$$0 = l^3 = d_4 l^2 = d_4 = c_4 lk = c_4 b_{43} = l^2 k = c_4 ,$$

so that there is no pure algebra in this case.*

[232^4]. The defining equation of this case is

$$k^2 = 0,$$

which gives

$$0 = lj = l^2 i = d_4 j = d_4 ,$$
$$0 = kl^2 = c_{34} kl = c_{34} , \quad kl = b_{34} j ,$$
$$0 = l^2 k = c_{43} lk = c_{43} , \quad lk = b_{43} j ,$$

and there can be no pure algebra if c_4 vanishes, so that it may be assumed, without loss of generality, that

$$l^2 = k ,$$

which gives

$$0 = l^3 = lk = kl .$$

This gives a quadruple algebra which may be called (r_4), its multiplication table being †

(r_4)	i	j	k	l
i	j	0	0	0
j	0	0	0	0
k	0	0	0	0
l	j	0	0	k

[24]. The defining equations of this case are

$$i^2 = j^2 = k^2 = l^2 = 0 ,$$

---

* Substituting $i - l$ for i, this case is, $i = B : D$, $j = A : D$, $k = A : C + C : D$, $l = A : B$.   [C. S. P.]
† $i = A : B + B : D + C : D$, $j = A : D$, $k = A : E$, $l = A : C + C : E$.   [C. S. P.]

and it may be assumed, from §§ 63 and 65, that

$$ij = k = -ji, \quad il = li = 0,$$

which give

$$0 = ik = ki = jk = kj = kl = lk,$$
$$0 = ijl = b_{24}k = b_{24} = j^2l = -a_{24}k + d_{21}jl = d_{24} = a_{21},$$
$$jl = -lj = c_{24}k,$$

so that there is no pure algebra in this case.*

QUINTUPLE ALGEBRA.

There are two cases:

> [1], when there is an idempotent basis;
> [2], when the algebra is nilpotent.

[1]. The defining equation of this case is

$$i^2 = i.$$

There are eleven cases:

[1²], when j, k, l and m are all in the first group;

[12], when j, k and l are in the first, and m in the second group;

[13], when j and k are in the first, and l and m in the second group;

[14], when j and k are in the first, l in the second, and m in the third group;

[15], when j is in the first, and k, l and m in the second group;

[16], when j is in the first, k and l in the second, and m in the third group;

[17], when j is in the first, k in the second, l in the third, and m in the fourth group;

[18], when j, k and l are in the second, and m in the third group;

[19], when j and k are in the second, and l and m in the third group;

[10¹], when j and k are in the second, l in the third, and m in the fourth group;

[11¹], when j is in the second, k in the third, and l and m in the fourth group.

[1²]. The defining equations of this case are

$$ij = ji = j, \quad ik = ki = k, \quad il = li = l, \quad im = mi = m.$$

The algebras deduced by §69 from algebras (l_4) to (r_4) may be named (a_5) to (j_5), and their multiplication tables are respectively

---

*$i = -A : C + B : E$, $j = A : B + C : E + cD : E$, $k = A : E$, $l = -A : D + cB : E$.   [C. S. P.]

(a_5)

	i	j	k	l	m
i	i	j	k	l	m
j	j	k	l	m	0
k	k	l	m	0	0
l	l	m	0	0	0
m	m	0	0	0	0

(b_5)

	i	j	k	l	m
i	i	j	k	l	m
j	j	k	l	0	0
k	k	l	0	0	0
l	l	0	0	0	0
m	m	l	0	0	l

(c_5)

	i	j	k	l	m
i	i	j	k	l	m
j	j	k	l	0	0
k	k	l	0	0	0
l	l	0	0	0	0
m	m	l	0	0	0

(d_5)

	i	j	k	l	m
i	i	j	k	l	m
j	j	k	0	0	l
k	k	0	0	0	0
l	l	0	0	0	0
m	m	$ak+bl$	0	0	l

(e_5) or

	i	j	k	l	m
i	i	j	k	l	m
j	j	k	0	0	l
k	k	0	0	0	0
l	l	0	0	0	0
m	m	$k-l$	0	0	0

(e_5)

	i	j	k	l	m
i	i	j	k	l	m
j	j	0	0	0	l
k	k	0	0	0	0
l	l	0	0	0	0
m	m	k	0	0	0

(f_5)	i	j	k	l	m
i	i	j	k	l	m
j	j	k	0	0	l
k	k	0	0	0	0
l	l	0	0	0	0
m	m	al	0	0	0

(g_5)	i	j	k	l	m
i	i	j	k	l	m
j	j	k	0	0	0
k	k	0	0	0	0
l	l	0	0	0	0
m	n	l	0	0	$l+ak$

(h_5)	i	j	k	l	m
i	i	j	k	l	m
j	j	k	0	0	0
k	k	0	0	0	0
l	l	0	0	0	0
m	m	l	0	0	k

(i_5)	i	j	k	l	m
i	i	j	k	l	m
j	j	k	0	0	0
k	k	0	0	0	0
l	l	0	0	0	0
m	m	l	0	0	0

(j_5)	i	j	k	l	m
i	i	j	k	l	m
j	j	k	0	0	0
k	k	0	0	0	0
l	l	0	0	0	0
m	m	k	0	0	l

[12]. The defining equations of this case are

$$ij = ji = j, \quad ik = ki = k, \quad il = li = l, \quad im = m, \quad mi = 0,$$

which give, by § 46,

$$0 = mj = mk = ml = m^2,$$

and if A is any expression belonging to the first group, but not involving i, we have the form

$$Am = am,$$

and by § 67, A is nilpotent, so that there is some power n which gives

$$0 = A^n = A^n m = aA^{n-1}m = a^n m = a = Am,$$
$$0 = jm = km = lm;$$

and there is no pure algebra in this case.*

[13]. The defining equations of this case are

$$ij = ji = j, \quad ik = ki = k, \quad il = l, \quad im = m, \quad li = mi = 0,$$

which give, by § 46,

$$0 = lj = lk = l^2 = lm = mj = mk = ml = m^2;$$

and it may be assumed from (a_3), by § 69, that

$$j^2 = k, \quad j^3 = 0.$$

It may also be assumed that

$$jl = m, \quad \text{whence} \dagger \quad kl = jm = 0.$$

We thus obtain a quintuple algebra which may be called (k_5), its multiplication table being this: ‡

---

*In fact i and m, by themselves, form the algebra (b_2), while i, j, k, l, by themselves form one of the algebras (a_4), (b_4), (c_4), (d_4), the products of m with j, k and l vanishing. [C. S. P.]

† This is proved as follows: $0 = j^3 l = j^2 m = d_{25} jl + e_{25} jm = d_{25} e_{25} l + (d_{25} + e_{25}^2) m$. Thus $d_{25} e_{25} = 0$ and $d_{25} + e_{25}^2 = 0$; or $d_{25} = 0$, $e_{25} = 0$, $jm = kl = 0$. [C. S. P.]

‡ $i = A:A + B:B + C:C$, $j = A:B + B:C$, $k = A:C$, $l = B:D$, $m = A:D$. [C. S. P.]

(k_5)	i	j	k	l	m
i	i	j	k	l	m
j	j	k	0	m	0
k	k	0	0	0	0
l	0	0	0	0	0
m	0	0	0	0	0

[14]. The defining equations of this case are

$$ij = ji = j, \quad ik = ki = k, \quad il = l, \quad mi = m, \quad li = im = 0,$$

which give, by § 46,

$$0 = jm = km = lj = lk = l^2 = ml = m^2.$$

It may be assumed from § 69 and (a_3) that

$$j^2 = k, \quad j^3 = 0,$$

whence

$$0 = jl = kl = mj = mk = jlm = a_{45}j + b_{45}k = a_{45} = b_{45}, \quad lm = c_{45}k,$$

and there is no pure algebra in this case.*

[15]. The defining equations of this case are

$$ij = ji = j, \quad ik = k, \quad il = l, \quad im = m, \quad ki = li = mi = 0,$$

which give, by §§ 46 and 67,

$$0 = j^2 = kj = k^2 = kl = km = lj = lk = l^2 = lm = mj = mk = ml = m^2.$$

It may be assumed that $\qquad jk = l, \quad jm = 0,$†

whence, $\qquad\qquad\qquad jl = 0,$

and there is no pure algebra in this case.‡

---

*$i = A : A + B : B + C : C$, $j = A : B + B : C$, $k = A : C$, $l = A : D$, $m = cD : C$. [C. S. P.]

† We cannot suppose $jk = k$, because $j^2 k = 0$. We may, therefore, put l for jk. Then $jl = 0$. Then, $0 = j^2 m = c_{25}e_{25}k + (d_{25}e_{25} + c_{25})l + e_{25}^2 m$. It follows that $jm = d_{25}l$, and substituting $m - d_{25}k$ for m, we have $jm = 0$. The algebra thus separates into (b_2) and (e_4). [C. S. P.]

‡ $i = A : A + B : B$, $j = A : B$, $k = B : C$, $l = A : C$, $m = A : D$. [C. S. P.]

[16]. The defining equations of this case are

$$ij = ji = j, \quad ik = k, \quad il = l, \quad mi = m, \quad ki = li = im = 0,$$

which give, by §§ 46 and 67,

$$0 = j^2 = jm = kj = k^2 = kl = lj = lk = l^2 = mj = mk = ml = m^2,$$
$$km = a_{35}i + b_{35}j, \quad lm = a_{45}i + b_{45}j,$$

and it may be assumed that

$$jk = d_{23}l, \quad jl = 0,$$

and d_{23} cannot vanish in the case of a pure algebra,* so that it is no loss of generality to assume

$$jk = l,$$

which gives

$$jkm = lm = a_{35}j.$$

There are two cases:

[161], when a_{35} does not vanish;
[162], when a_{35} vanishes.

[161]. The defining equation of this case can be reduced to

$$a_{35} = 1,$$

which gives

$$lm = j, \quad km = i + b_{35}j,$$

and $i + b_{35}j$ can be substituted for i, and this gives a quintuple algebra which may be called (l_5), of which the multiplication table is

(l_5)	i	j	k	l	m
i	i	j	k	l	
j	i	0	l	0	0
k	0	0	0	0	i
l	0	0	0	0	j
m	m	0	0	0	0

---

* But $0 = mk = kmk = (a_{35}i + b_{35}j)k = a_{35}k + d_{23}b_{35}l$. Hence $a_{35} = 0$ and either d_{23} or $b_{35} = 0$, and in either case there is no pure algebra. The two algebras (l_5) and (m_5) are incorrect, as may be seen by comparing $k \cdot mk$ with $km \cdot k$. [C. S. P.]

[162]. The defining equation of this case is

$$a_{35} = 0,$$

which gives

$$km = b_{35}j, \quad lm = 0;$$

and b_{35} cannot vanish in the case of a pure algebra, so that it is no loss of generality to assume

$$km = j.$$

This gives a quintuple algebra which may be called (m_5), of which the multiplication table is

(m_5)	i	j	k	l	m
i	i	j	k	l	0
j	j	0	l	0	0
k	0	0	0	0	j
l	0	0	0	0	0
m	m	0	0	0	0

[17]. The defining equations of this case are

$$ij = ji = j, \quad ik = k, \quad li = l, \quad ki = il = im = mi = 0,$$

which give, by §§ 46 and 67,

$$0 = j^2 = jk = jl = jm = kj = k^2 = lj = l^2 = lm = mj = mk,$$
$$kl = a_{34}i + b_{34}j, \quad km = c_{35}k, \quad lk = c_{43}m, \quad ml = d_{54}l, \quad m^2 = e_5 m,$$
$$0 = jkl = a_{34}j = a_{34},$$
$$lkl = b_{34}lj = 0 = c_{43}ml = c_{43}d_{54}, \quad klk = b_{34}jk = 0 = a_{43}km - c_{43}c_{35},$$
$$lkm = c_{35}lk = c_{43}m^2 = c_{35}c_{43}m = 0 = c_{43}e_5,$$
$$kml = d_{54}kl = c_{35}kl, \quad (d_{54} - c_{35})b_{34} = 0, \quad km^2 = e_5 km = c_{35}km, \quad (e_5 - c_{35})c_{35} = 0,$$
$$m^2l = e_5 ml = d_{54}ml, \quad (e_5 - d_{54})d_{54} = 0.$$

There are two cases:

$$[171], \text{ when } e_5 = 1 \, ;^*$$
$$[172], \text{ when } e_5 = 0 \, .$$

[171]. The defining equation of this case is

$$m^2 = m \, ,$$

which gives

$$0 = c_{43} = lk \, .$$

There can be no pure algebra if either of the quantities b_{34}, c_{35} or d_{54} vanish, and there is no loss of generality in assuming

$$kl = j \, , \quad km = k \, , \quad ml = l \, .$$

This gives a quintuple algebra which may be called (n_5), its multiplication table being

(n_5)	i	j	k	l	m
i	i	j	k	0	0
j	j	0	0	0	0
k	0	0	0	j	k
l	l	0	0	0	0
m	0	0	0	l	m

[172]. The defining equation of this case is

$$m^2 = 0 \, ,$$

which gives

$$0 = c_{35} = d_{54} = km = ml \, ;$$

---

* But on examination of the assumptions already made, it will be seen that if e_5 is not zero, and consequently $c_{43} = 0$, the algebra breaks up into two. Accordingly, the algebra (n_5) is impure, for i, j, k and l, alone, form the algebra (f_4), while m, l, k, j, alone, form the algebra (h_4), and $im = mi = 0$. [C. S. P.]

and there can be no pure algebra if either b_{34} or c_{43} vanishes, and it may be assumed that

$$kl = j, \quad lk = m.$$

This gives a quintuple algebra which may be called (o_5), its multiplication table being as follows : *

(o_5)	i	j	k	l	m
i	i	j	k	0	0
j	j	0	0	0	0
k	0	0	0	j	0
l	l	0	m	0	0
m	0	0	0	0	0

[18]. The defining equations of this case are

$$ij = j, \quad ik = k, \quad il = l, \quad mi = m, \quad ji = ki = li = im = 0,$$

which give, by § 46,

$$0 = j^2 = jk = jl = kj = k^2 = kl = lj = lk = l^2 = mj = mk = ml = m^2.$$

But if A is any expression of the second group,

$$Am = ai;$$

which gives

$$0 = Amj = aj = a = Am = jm = km = lm,$$

and there is no pure algebra in this case.

[19]. The defining equations of this case are

$$ij = j, \quad ik = k, \quad li = l, \quad mi = m, \quad il = im = ji = ki = 0,$$

which give, by § 46,

$$0 = j^2 = jk = kj = k^2 = lj = lk = l^2 = lm = mj = mk = ml = m^2.$$

---

* $i = R:B + D:D + F:F$, $j = D:F$, $k = B:C + D:E$, $l = A:B + E:F$, $m = A:C$. [C. S. P.]

But if A is an expression of the second group and B one of the third,

$$AB = ai,$$

which gives

$$0 = ABj = aj = a = AB = jl = jm = kl = lm,$$

and there is no pure algebra in this case.

[10′]. The defining equations of this case are

$$ij = j, \quad ik = k, \quad li = l, \quad ji = ki = il = im = mi = 0,$$

which give, by § 46,

$$0 = j^2 = jk = kj = k^2 = l^2 = lm = mj = mk;$$

and it is obvious that we may assume

$$jl = 0.$$

We have, then,

$$jm = b_{25}j + c_{25}k, \quad kl = a_{34}i, \quad km = b_{35}j + c_{35}k,$$
$$lj = e_{42}m, \quad lk = e_{43}m, \quad ml = d_{54}l, \quad m^2 = e_5 m,$$
$$0 = d_{54}jl = jml = c_{25}kl = a_{34}c_{25}i = a_{34}c_{25}.$$

There are two cases:

[10′1], when a_{34} does not vanish;
[10′2], when a_{34} vanishes.

[10′1]. The defining equation of this case can be reduced to

$$kl = i,$$

which gives

$$c_{25} = 0, \quad jm = b_{25}j.$$

There are two cases:

[10′1²], when $e_5 = 1$;
[10′12], when e_5 vanishes.

[10′1²]. The defining equation of this case is

$$m^2 = m;$$

and we assume

$$jm = j, \quad ml = l, \quad km = k,$$

because otherwise this case would coincide with a subsequent one. We get, then,

$$0 = jlj = e_{42}jm = e_{42} = lj, \qquad 0 = jlk = e_{43}jm = e_{43} = lk,$$

which virtually brings this case under [10′2].*

---

* This does not seem clear. But $i = i^2 = klkl = 0$, which is absurd. [C. S. P.]

[10'12]. The defining equation of this case is

$$m^2 = 0,$$

which gives

$$0 = jm^2 = b_{25}jm = b_{25} = jm, \quad 0 = m^2l = d_{52}ml = a_{54} = ml,$$
$$0 = km^2 = c_{35}km = c_{35}, \quad km = b_{35}j, \quad lkl = li = l = c_{43}ml = 0,$$

which is impossible, and this case disappears.

[10'2]. The defining equation of this case is

$$kl = 0.*$$

There are two cases:

[10'21], when $e_5 = 1$;

[10'2²], when e_5 vanishes.

[10'21]. The defining equation of this case is

$$m^2 = m,$$

and if we would not virtually proceed to a subsequent case, we must assume

$$jm = j, \quad km = k, \quad ml = l,$$

and there is no loss of generality in assuming

$$lj = 0,$$

so that there is no pure algebra in this case.†

[10'2²]. The defining equation of this case is

$$m^2 = 0,$$

which gives

$$0 = m^2l = d_{54}ml = d_{54} = ml;$$

and we may assume

$$c_{25} = 0,$$

which gives

$$0 = jm^2 = b_{25}jm = b_{25} = jm, \quad 0 = km^2 = c_{35}km = c_{35}, \quad km = b_{35}j,$$
$$0 = e_{43}m^2 = lkm = b_{35}e_{42}m = b_{35}e_{42};‡$$

---

* In this case, the algebra at once separates into an algebra between j, k, l and m, and three double algebras between i and j, i and k, and i and l, respectively. [C. S. P.]

† In fact, $0 = lklk = e_{43}^2m = e_{43} = lk$. So that the algebra falls into six parts of the form (b_2). [C. S. P.]

‡ The author omits to notice that $0 = klk = e_{43}km = e_{43}b_{35}$. Thus, either $km = 0$ or $lj = lk = 0$. The algebra (p_5) involves an inconsistency in regard to klk. [C. S. P.]

and we have without loss of generality

$$lj = 0, \quad km = j, \quad lk = m.$$

This gives a quintuple algebra which may be called (p_5), of which the multiplication table is

(p_5)	i	j	k	l	m
i	i	j	k	0	0
j	0	0	0	0	0
k	0	0	0	0	j
l	l	0	m	0	0
m	0	0	0	0	0

[11′]. The defining equations of this case are

$$ij = j, \quad ki = k, \quad ji = ik = il = im = li = mi = 0;$$

which give, by § 46,

$$0 = j^2 = k^2 = kl = km = lj = mj,$$

$$jk = a_{23}i, \quad jl = b_{24}j, \quad jm = b_{25}j, \quad kj = d_{32}l + e_{32}m, \quad lk = c_{43}k, \quad mk = c_{53}k.$$

There are two cases:

[11′1], when l is the idempotent base of the fourth group;
[11′2], when the fourth group is nilpotent.

[11′1]. The defining equation of this case is

$$l^2 = l.$$

There are two cases:

[11′1²], when m is in the second subsidiary group of the fourth group;
[11′12], when m is in the fourth subsidiary group of the fourth group.

[11′1²]. The defining equations of this case are

$$lm = m, \quad ml = 0;$$

which give
$$0 = m^2 = jm^2 = b_{25}jm = b_{25} = jm,$$
$$0 = m^2k = c_{53}mk = c_{53} = mk;$$

and a_{23} cannot vanish in a pure algebra, so that we may assume

$$jk = i,$$

which gives

$$kjk = k = d_{32}c_{43}k, \quad jkj = j = d_{32}b_{24}j, \quad 1 = d_{32}c_{43} = d_{32}b_{24},$$
$$jl = jl^2 = b_{24}jl, \quad b_{24}^2 = b_{24} = 1, \quad lk = l^2k = c_{43}lk, \quad c_{43}^2 = c_{43} = 1 = d_{32},$$
$$jl = j, \quad lk = k, \quad kjl = l = kj,$$

and there is no pure algebra in this case.*

[11′12]. The defining equations of this case are

$$lm = ml = 0,$$

which give

$$0 = jlm = b_{24}jm = b_{24}b_{25}j = b_{24}b_{25}, \quad 0 = lmk = c_{53}lk = c_{43}c_{53}k = c_{43}c_{53},$$
$$kjl = d_{32}l = b_{24}kj = b_{24}d_{32}l + b_{24}e_{32}m, \quad lkj = d_{32}l = c_{43}kj = c_{43}d_{32}l + c_{43}e_{32}m,$$
$$kjm = e_{32}m^2 = b_{25}kj = b_{25}d_{32}l + b_{25}e_{32}m, \quad mkj = e_{32}m^2 = c_{53}kj = c_{53}d_{32}l + c_{53}e_{32}m,$$
$$d_{32} = b_{24}d_{32} = c_{43}d_{32}, \quad 0 = b_{25}d_{32} = e_{53}d_{32} = b_{24}e_{32} = c_{43}e_{32}.†$$

There are two cases:
$$[11′121], \text{ when } m \text{ is idempotent};$$
$$[11′12^2], \text{ when } m \text{ is nilpotent}.$$

[11′121]. The defining equation of this case is

$$m^2 = m,$$

which gives

$$e_{32} = c_{53}e_{32} = b_{25}e_{32};$$

and it may be assumed that

$$b_{24} = 0.$$

But if the algebra is then regarded as having l for its idempotent basis, it is evident from § 50 that the bonds required for a pure algebra are wanting, so that there is no pure algebra in this case.‡

---

* In fact, i, j, k, l form the algebra (g_4), and l, m, the algebra (b_2) [C S P]

† The last equation holds by ¿ 68. [C. S. P.]

‡ Namely, $d_{32} = 0$, and either $e_{32} = 1$, when l forms the algebra (a_1), and i, j, k, m the algebra (g_4), or else $e_{32} = 0$, when by [13] of triple algebra $a_{23} = 0$, and j and k each forms the algebra (b_2) with each of the letters i, l, m. [C. S. P.]

[11′12²]. The defining equation of this case is

$$m^2 = 0,$$

which gives

$$0 = jm^2 = b_{25}jm = b_{25}^2 j = b_{25} = jm, \quad 0 = m^2 k = c_{53}mk = c_{53}^2 k = c_{53} = mk,$$
$$1 = b_{24} = c_{43}, \quad jl = j, \quad lk = k, \quad 0 = e_{32},$$

and there is no pure algebra in this case.*

[11′2]. The defining equation of this case is

$$l^n = 0,$$

in which n is 2 or 3. We must then have

$$0 = lm = ml = m^2,$$

which give

$$0 = jl^3 = b_{24}jl^2 = b_{24}^2 jl = b_{24} = jl = jm = lk = mk, \quad 0 = kjk = a_{23}k = a_{23} = jk,$$

and there is no pure algebra in this case. †

[2]. The defining equation of this case is

$$i^n = 0.$$

There are five cases:

     [21], when $n = 6$;
     [2²], when $n = 5$;
     [23], when $n = 4$;
     [24], when $n = 3$;
     [25], when $n = 2$.

[21]. The defining equation of this case is

$$i^6 = 0,$$

and by § 60,

$$i^2 = j, \quad i^3 = k, \quad i^4 = l, \quad i^5 = m.$$

This gives a quintuple algebra which may be called (q_5), its multiplication table being

---

(q_5)	i	j	k	l	m
i	j	k	l	m	0
j	k	l	m	0	0
k	l	m	0	0	0
l	m	0	0	0	0
m	0	0	0	0	0

[2^3]. The defining equation of this case is

$$i^5 = 0,$$

and by § 59,

$$i^2 = j, \quad i^3 = k, \quad i^4 = l.$$

There are then by § 64 two quintuple algebras which may be called (r_5) and (s_5), their multiplication tables being

(r_5)	i	j	k	l	m
i	j	k	l	0	0
j	k	l	0	0	0
k	l	0	0	0	0
l	0	0	0	0	0
m	l	0	0	0	l

(s_5)	i	j	k	l	m
i	j	k	l	0	0
j	k	l	0	0	0
k	l	0	0	0	0
l	0	0	0	0	0
m	l	0	0	0	0

[23]. The defining equation of this case is

$$i^4 = 0;$$

and by § 59,

$$i^2 = j, \quad i^3 = k;$$

and it may be assumed, from the principle of § 63, that

$$il = 0,$$

which gives

$$0 = jl = kl = ili = il^2 = ilm$$

$$li = c_{41}k + d_{41}l + e_{41}m, \quad l^2 = c_4k + d_4l + e_4m, \quad lm = c_{45}k + d_{45}l + e_{45}m.$$

There are two cases:

[231], when $im = l$;

[232], when $im = 0$.

[231]. The defining equation of this case is

$$im = l,$$

whence

$$0 = jm = km = jmi = jml = jm^2 = e_{41} = e_4 = e_{45},$$

$$li^2 = d_4li, \quad 0 = li^4 = d_{41}li^3 = d_{41}^2li^2 = d_{41}^3li = d_{41} = lj = lk,$$

$$l^3 = d_4l^2, \quad 0 = l^4 = d_4l^3 = d_4, \quad li = c_{41}k, \quad l^2 = c_4k, \quad lm = c_{45}k + d_{45}l,$$

$$imi = li = c_{41}k, \quad mi = c_{41}j + c_{51}k + d_{51}l, \quad mj = c_{41}(1 + d_{51})k, \quad mk = 0,$$

$$iml = l^2 = c_4k, \quad ml = c_4j + c_{54}k + d_{54}l,$$

$$im^2 = lm = c_{45}k + d_{45}l, \quad m^2 = c_{45}j + c_5k + d_5l + d_{45}m,$$

$$0 = m^4 = d_{45}, \quad lim = l^2 = c_{41}km = 0 = mli = d_{54}c_{41}k = d_{54}c_{41},$$

$$0 = mlm = d_{54}lm = d_{54}c_{45}, \quad 0 = m^2l = d_{54}ml = d_{54}.^{*}$$

There are two cases:

[231²], when c_{41} does not vanish;

[2312], when c_{41} vanishes.

[231²]. The defining equation of this case is reducible to

$$li = k.$$

There are two cases:

[231³], when c_{45} does not vanish;

[231²2], when c_{45} vanishes.

[231³]. The defining equation of this case can be reduced to

$$lm = k,$$

which gives

$$m^2i = k + d_{51}k + d_{51}^2k = k + d_5k, \quad d_5 = d_{51} + d_{51}^2,$$

$$m^3 = k + d_{51}k + d_5d_{51}k = d_5k, \quad d_{51}^3 = -1;$$

---

* To these equations are to be added the following, which is taken for granted below: $ml \doteq mim = c_{45}d_{51}k$. [C. S. P.]

and if $\mathfrak{r}$ is one of the imaginary cube roots of -1, there are two cases:

$$[231^4], \quad \text{when } d_{51} = \mathfrak{r};$$
$$[231^32], \quad \text{when } d_{51} = -1.$$

$[231^4]$. The defining equation of this case is

$$d_{51} = \mathfrak{r},$$

which gives

$$i(m - c_{51}l) = l, \quad l(m - c_{51}l) = k,$$
$$(m - c_{51}l)i = j + \mathfrak{r}l, \quad (m - c_{51}l)j = (1 + \mathfrak{r})k,$$
$$(m - c_{51}l)k = 0, \quad (m - c_{51}l)l = \mathfrak{r}k,$$
$$(m - c_{51}l)^2 = j + [c_5 - c_{51}(1 + \mathfrak{r})]k + (2\mathfrak{r} - 1)l;$$

so that the substitution of $m - c_{51}l$ for m is the same as to make

$$c_{51} = 0.$$

There are two cases:

$$[231^5], \quad \text{when } c_5 \text{ does not vanish};$$
$$[231^42], \quad \text{when } c_5 \text{ vanishes.}$$

$[231^5]$. The defining equation of this case can be reduced to

$$c_5 = 1.$$

There is then a quintuple algebra which may be called (t_5), its multiplication table being *

---

* The author has overlooked the circumstance that (t_5) and (u_5) are forms of the same algebra. If in (t_5) we put $i_1 = i - \mathfrak{r}^2 j$, $j_1 = j - 2\mathfrak{r}^2 k$, $k_1 = k$, $l_1 = -\mathfrak{r}^2 k + l$, $m_1 = -\mathfrak{r}^2 j + m$, we get (u_5). The structure of this algebra may be shown by putting $i_1 = \mathfrak{r}i$, $j_1 = \mathfrak{r}^2 j$, $k_1 = -k$, $l_1 = \mathfrak{r}^2 j - \mathfrak{r}l$, $m_1 = \mathfrak{r}i - m$, when we have this multiplication table (where the subscripts are dropped):

(u_5)	i	j	k	l	m
i	j	k	0	k	l
j	k	0	0	0	k
k	0	0	0	0	0
l	$\mathfrak{r}k$	0	0	0	0
m	$\mathfrak{r}l$	$\mathfrak{r}^2 k$	0	0	0

In relative form, $i = A : B + A : C + B : E + C : D + E : G$, $j = A : D + A : E + B : G$, $k = A : G$, $l = \mathfrak{r}A : E + C : G$, $m = \mathfrak{r}^2 A : B + A : F + \mathfrak{r}C : E + D : G - F : G$. [C. S. P.]

(t_5)	i	j	k	l	m
i	j	k	0	0	l
j	k	0	0	0	0
k	0	0	0	0	0
l	k	0	0	0	k
m	$j+\mathfrak{r}l$	$(1+\mathfrak{r})k$	0	$\mathfrak{r}k$	$j+k+(2\mathfrak{r}-1)l$

[231^{4}2]. The defining equation of this case is

$$c_5 = 0.$$

There is then a quintuple algebra which may be called (u_5), its multiplication table being

(u_5)	i	j	k	l	m
i	j	k	0	0	l
j	k	0	0	0	0
k	0	0	0	0	0
l	k	0	0	0	k
m	$j+\mathfrak{r}l$	$(1+\mathfrak{r})k$	0	$\mathfrak{r}k$	$j+(2\mathfrak{r}-1)l$

[231^{3}2]. The defining equation of this case is

$$d_{51} = -1,$$

which gives

$$d_5 = 0, \quad i(m-c_{51}l) = l, \quad l(m-c_{51}l) = k,$$
$$(m-c_{51}l)i = j-l, \quad (m-c_{51}l)l = -k, \quad (m-c_{51}l)_2 = j+c_5k;$$

so that the substitution of $m - c_{51}l$ * for m is the same as to make

$$c_{51} = 0.$$

---

* The original text has $m-c_{51}k$ throughout these equations, but it is plain that $m-c_{51}l$ is meant. [C. S. P.]

There are two cases :

$[231^3 21]$, when c_5 does not vanish;

$[231^3 2^2]$, when c_5 vanishes.

$[231^3 21]$. The defining equation of this case can be reduced to

$$c_5 = 1.$$

There is a quintuple algebra which may be called (v_5), its multiplication table being *

(v_5)	i	j	k	l	m
i	j	k	0	0	l
j	k	0	0	0	0
k	0	0	0	0	0
l	k	0	0	0	k
m	$j-l$	0	0	$-k$	$j+k$

$[231^3 2^2]$. The defining equation of this case is

$$c_5 = 0.$$

This gives a quintuple algebra which may be called (w_5), its multiplication table being *

---

* The algebra (v_5) reduces to (w_5) on substituting $i_1 = i + \frac{1}{3}j + \frac{1}{3}l$, $j_1 = j + k$, $k_1 = k$, $l_1 = \frac{2}{3}k + l$, $m_1 = \frac{1}{3}j + \frac{1}{3}l + m$. To exhibit the structure of this algebra, we may put ρ and ρ' for imaginary cube roots of 1, and substitute in (w_5) $i_1 = i + \rho'm$, $j_1 = (1-\rho)j + k + \sqrt{-3}l$, $k_1 = 3k$, $l_1 = (1-\rho')j + k - \sqrt{-3}l$, $m_1 = i + \rho m$. Then, dropping the subscripts, we have this multiplication table.

	i	j	k	l	m
i	0	0	0	k	j
j	l	0	0	0	0
k	0	0	0	0	0
l	0	0	0	0	k
m	l	k	0	0	0

In relative form, $i = \rho'A : B + \rho'C : F + 3\rho D : E$, $j = 3\rho A : C + 3\rho'D : F$, $k = 3A : D$, $l = 3\rho'A : E + 3\rho B : F$. $m = \rho A : D + 3\rho'B : C + \rho E : F$. [C. S. P.]

(w)	i	j	k	l	m
i	j	k	0	0	l
j	k	0	0	0	0
k	0	0	0	0	0
l	k	0	0	0	k
m	$j-l$	0	0	$-k$	$j+k$

[$231^2 2$]. The defining equation of this case is

$$lm = 0 ,$$

which gives

$$ml = 0 , \quad m^2 = c_5 k + d_5 l , \quad m^2 i = d_5 k = [1 + d_{51}] k , \quad d_5 = 1 + d_{51} ,$$

and c_{51} may be made to vanish without loss of generality.

There are three cases:

[$231^2 21$], when neither d_{51} nor $d_{51} + 1$ vanishes;
[$231^2 2^2$], when $d_{51} + 1$ vanishes;
[$231^2 23$], when d_{51} vanishes.

[$231^2 21$]. The defining formulae of this case are

$$d_{51} \gtrless 0 , \quad d_{51} \gtrless -1 .$$

There are two cases:

[$231^2 21^2$], when c_5 does not vanish;
[$231^2 212$], when c_5 vanishes.

[$231^2 21^2$]. The defining equation of this case can always be reduced to

$$c_5 = 1 .$$

This gives a quintuple algebra which may be called (x_5), its multiplication table being*

---

* In relative form, $i = A : B + A : E + B : D + D : F$, $j = A : D + B : F$, $k = A : F$, $l = A : D$, $m = (1 + a) A : B + A : C + A : E + B : D + C : D + D : F + E : F$. [C. S. P.]

(x_5)	i	j	k	l	m
i	j	k	0	0	l
j	k	0	0	0	0
k	0	0	0	0	0
l	k	0	0	0	0
m	$j+al$	$(1+a)k$	0	0	$k+(1+a)l$

[231^2212]. The defining equation of this case is

$$c_5 = 0.$$

This gives a quintuple algebra which may be called (y_5), its multiplication table being *

(y_5)	i	j	k	l	m
i	j	k	0	0	l
j	k	0	0	0	0
k	0	0	0	0	0
l	k	0	0	0	0
m	$j+al$	$(1+a)k$	0	0	$(1+a)l$

[231^22^2]. The defining equation of this case is

$$d_{51} = -1,$$

which gives

$$mi = j - l, \quad mj = 0, \quad m^2 = c_5 k.$$

There are two cases:

[231^22^21], when c_5 does not vanish;
[231^22^3], when c_5 vanishes.

---

* The relative form is the same as that of (x_5); omitting from m the terms $A : E$ and $E : F$. [C. S. P.]

[$231^2 2^2 1$]. The defining equation of this case can be reduced to

$$m^2 = k.$$

This gives a quintuple algebra which may be called (z_5), its multiplication table being *

(z_5)	i	j	k	l	m
i	j	k	0	0	l
j	k	0	0	0	0
k	0	0	0	0	0
l	k	0	0	0	0
m	$j-l$	0	0	0	k

[$231^2 2^3$]. The defining equation of this case is

$$m^2 = 0.$$

This gives a quintuple algebra which may be called (aa_5), its multiplication table being †

(aa_5)	i	j	k	l	m
i	j	k	0	0	l
j	k	0	0	0	0
k	0	0	0	0	0
l	k	0	0	0	0
m	$j-l$	0	0	0	0

---

*In relative form, $i = A:B + B:C + C:D$, $j = A:C + B:D$, $k = A:D$, $l = A:C$, $m = B:C$ $+ A:E + E:D$. [C. S. P.]

† In relative form, the same as (z_5), except that $m = B:C$. [C. S. P.]

[$231^2 23$]. The defining equation of this case is

$$mi = j,$$

which gives

$$0 = (l-j)i = (m-i)i;$$

so that, by the substitution of $l-j$ for l and $m-i$ for m, this case would virtually be reduced to [232].

[2312]. The defining equation of this case is

$$li = 0,$$

which gives

$$mj \doteq 0, \quad mim = ml = d_{51}lm, \quad d_{45} = 0, \quad c_{54} = d_{51}c_{45},$$
$$m^2i = d_{51}ml = c_{45}k, \quad c_{45} = d_{51}c_{54}, \quad m^3 = d_5lm = d_5ml, \quad d_5(c_{54} - c_{45}) = 0.$$

There are two cases:

[23121], when d_5 does not vanish;

[2312^2], when d_5 vanishes.

[23121]. The defining equation of this case can be reduced to

$$d_5 = 1,$$

which gives

$$c_{45} = c_{54};$$

and it may be assumed without loss of generality that

$$c_5 = 0.*$$

There are two cases:

[23121^2], when c_{45} does not vanish;

[231212], when c_{45} vanishes.

[23121^2]. The defining equation of this case can be reduced to

$$lm = ml = k,$$

which gives

$$d_{51} = 1.$$

There are two cases:

[23121^3], when c_{51} does not vanish;

[$23121^2 2$], when c_{51} vanishes.

[23121^3]. The defining equation of this case can be reduced to

$$c_{51} = 1.$$

* Namely, by putting $l_1 = c_5 k + l$, $m_1 = m - c_5 j$. [C. S. P.]

This gives a quintuple algebra which may be called (ab_5), its multiplication table being *

(ab_5)	i	j	k	l	m
i	j	k	0	0	l
j	k	0	0	0	0
k	0	0	0	0	0
l	0	0	0	0	k
m	$k+l$	0	0	k	$j+l$

[23121²2]. The defining equation of this case is

$$c_{51} = 0.$$

This gives a quintuple algebra which may be called (ac_5), its multiplication table being †

---

* The structure of this algebra is best seen on making the following substitutions: Let $\mathfrak{h}_1$ and $\mathfrak{h}_2$ represent the two roots of the equation $x^2 = x+1$. That is, $\mathfrak{h}_1 = \frac{1}{2}(1+\sqrt{5})$ and $\mathfrak{h}_2 = \frac{1}{2}(1-\sqrt{5})$. Then substitute $i_1 = \mathfrak{h}_1^{-\frac{1}{3}}(i+\mathfrak{h}_1 m)$, $j_1 = \mathfrak{h}_1^{-\frac{4}{3}}\{(2+\mathfrak{h}_1)j+\mathfrak{h}_1 k+(1+3\mathfrak{h}_1)l\}$, $k_1 = \frac{1}{5}k$, $l_1 = \mathfrak{h}_2^{-\frac{4}{3}}\{(2+\mathfrak{h}_2)j + \mathfrak{h}_2 k + (1+3\mathfrak{h}_2)l\}$, $m_1 = \mathfrak{h}_2^{-\frac{1}{3}}(i+\mathfrak{h}_2 m)$. Then, we have the multiplication table:

	i	j	k	l	m
i	j	k	0	0	$\frac{1}{5}\mathfrak{h}_2 k$
j	k	0	0	0	0
k	0	0	0	0	0
l	0	0	0	0	k
m	$\frac{1}{5}\mathfrak{h}_1 k$	0	0	k	l

In relative form, $i = A:B+B:C+C:D+\frac{1}{5}\mathfrak{h}_1 A:G+ H:D$, $j = A:C+B:D$, $k=A:D$, $l=A:F +E:D$, $m=A:E+E:F+F:D+\frac{1}{5}\mathfrak{h}_2 A:H+G:D$. [C. S. P.]

† On making the same substitutions for i and m as in the last note, this algebra falls apart into two algebras of the form (b_3). [C. S. P.]

(ac_5)	i	j	k	l	m
i	j	k	0	0	l
j	k	0	0	0	0
k	0	0	0	0	0
l	0	0	0	0	k
m	l	0	0	k	$j+l$

[231212]. The defining equation of this case is

$$ml = lm = 0.$$

There are two cases :

[2312121], when c_{51} does not vanish ;

[231212²], when c_{51} vanishes.

[2312121]. The defining equation of this case can be reduced to

$$c_{51} = 1.$$

This gives a quintuple algebra which may be called (ad_5), its multiplication table being *

(ad_5)	i	j	k	l	m
i	j	k	0	0	l
j	k	0	0	0	0
k	0	0	0	0	0
l	0	0	0	0	0
m	$k+al$	0	0	0	l

---

* In relative form, $i = A : B + B : C + C : D + E : F + aF : G$, $j = A : C + B : D + aE : G$, $k = A : D$, $l = E : G$, $m = A : C + E : F + F : G$. [C. S. P.]

[231212^2]. The defining equation of this case is

$$c_{51} = 0.$$

This gives a quintuple algebra which may be called (ae_5), its multiplication table being *

(ae_5)	i	j	k	l	m
i	j	k	0	0	l
j	k	0	0	0	0
k	0	0	0	0	0
l	0	0	0	0	0
m	al	0	0	0	l

[2312^2]. The defining equation of this case is

$$d_5 = 0.$$

There are two cases :

[2312^{2}1], when c_{45} does not vanish ;
[2312^3], when c_{45} vanishes.

[2312^{2}1]. The defining equation of this case can be reduced to

$$lm = k,$$

which gives

$$c_{45} = d_{51}^2 c_{45}, \quad d_{51}^2 = 1.$$

There are two cases :

[2312^{2}1^2], when $d_{51} = 1$;
[2312^{2}12], when $d_{51} = -1$.

[2312^{2}1^2]. The defining equation of this case is

$$d_{51} = 1,$$

which gives

$$c_{54} = 1, \quad ml = k.$$

---

* In relative form, the same as (ad_5) except that $m = E : F + F : G$.   [C. S. P.]

There are two cases:

$$[2312^21^3], \quad \text{when } c_{51} \text{ does not vanish;}$$
$$[2312^21^22], \quad \text{when } c_{51} \text{ vanishes.}$$

$[2312^21^3]$. The defining equation of this case can be reduced to

$$c_{51} = 1.$$

This gives a quintuple algebra which may be called (af_5), its multiplication table being*

(af_5)	i	j	k	l	m
i	j	k	0	0	l
j	k	0	0	0	0
k	0	0	0	0	0
l	0	0	0	0	k
m	$k+l$	0	0	k	$j+ck$

---

* To show the construction of this algebra, we may substitute $i_1 = i + m$, $j_1 = 2j + (a+1)k + 2l$, $k_1 = 4k$, $l_1 = 2j + (a-1)k - 2l$, $m_1 = i - m$. This gives the following multiplication table:

	i	j	k	l	m
i	j	k	0	0	$-\dfrac{a-1}{4}k$
j	k	0	0	0	0
k	0	0	0	0	0
l	0	0	0	0	k
m	$-\dfrac{a+1}{4}k$	0	0	k	l

This algebra thus strongly resembles (ab_5). In relative form, $i = A : B + B : C + C : D + A : G - \dfrac{a+1}{4} G : D$, $j = A : C + B : D - \dfrac{a+1}{4} A : D$, $k = A : D$, $l = A : F + E : D - \dfrac{a-1}{4} A : D$, $m = A : E + E : F + F : D + A : G - \dfrac{a-1}{4} G : D$. [C. S. P.]

$[2312^21^22]$. The defining equation of this case is

$$c_{51} = 0.$$

There are two cases:

$[2312^21^221]$, when c_5 does not vanish;
$[2312^21^22^2]$, when c_5 vanishes.

$[2312^21^221]$. The defining equation of this case can be reduced to

$$c_5 = 1.$$

This gives a quintuple algebra which may be called (ag_5), its multiplication table being *

(ag_5)	i	j	k	l	m
i	j	k	0	0	l
j	k	0	0	0	0
k	0	0	0	0	0
l	0	0	0	0	k
m	l	0	0	k	$j+k$

$[2312^21^22^2]$. The defining equation of this case is

$$c_5 = 0.$$

This gives a quintuple algebra which may be called (ah_5), its multiplication table being †

---

* On substituting $i_1 = i + \frac{1}{2}j + m$, $m_1 = i + \frac{1}{2}j - m$, this algebra falls apart into two of the form (b_3). [C. S. P.]

† On substituting $i_1 = i + m$, $m_1 = i - m$, $j_1 = j + l$, $l_1 = j - l$, this algebra falls apart into two of the form (b_3). [C. S. P.]

(ah_5)	i	j	k	l	m
i	j	k	0	0	l
j	k	0	0	0	0
k	0	0	0	0	0
l	0	0	0	0	k
m	l	0	0	k	j

[$2312^2 12$]. The defining equation of this case is

$$d_{51} = -1,$$

which gives

$$c_{54} = -1, \quad ml = -k.$$

There are two cases:

[$2312^2 121$], when c_{51} does not vanish;
[$2312^2 12^2$], when c_{51} vanishes.

[$2312^2 121$]. The defining equation of this case can be reduced to

$$c_{51} = 1.$$

This gives a quintuple algebra which may be called (ai_5), its multiplication table being *

(ai_5)	i	j	k	l	$'m$
i	j	k	0	0	l
j	k	0	0	0	0
k	0	0	0	0	0
l	0	0	0	0	k
m	$k-l$	0	0	$-k$	$j+ck$

---

* In relative form, $i = A:C - B:F + C:E + D:G + E:G$, $j = A:E + C:G$, $k = A:G$, $= A:F - B:G$, $m = A:B + A:D + B:E + C:F + aD:G + F:G$. [C. S. P.]

[2312²12²]. The defining equation of this case is

$$mi = -l.$$

There are two cases:

[2312²12²1], when c_5 does not vanish;
[2312²12³], when c_5 vanishes.

[2312²12²1]. The defining equation of this case can be reduced to

$$c_5 = 1.$$

This gives a quintuple algebra which may be called (aj_5), its multiplication table being *

(aj_5)	i	j	k	l	m
i	j	k	0	0	l
j	k	0	0	0	0
k	0	0	0	0	0
l	0	0	0	0	k
m	$-l$	0	0	$-k$	$j+k$

[2312²12³]. The defining equation of this case is

$$m^2 = j.$$

This gives a quintuple algebra which may be called (ak_5), its multiplication table being †

---

* In relative form, $i = A:C + C:E + E:G - B:F$, $j = A:E + C:G$, $k = A:G$, $l = A:F - B:G$, $m = A:B + B:E + C:F + F:G + A:D + D:G$. [C. S. P.]

† In relative form, $i = A:C + C:D + D:F - B:E$, $j = A:D + C:F$, $k = A:F$, $l = A:E - B:F$, $m = A:B + B:D + C:E + E:F$. [C. S. P.]

(ak_5)	i	j	k	l	m
i	j	k	0	0	l
j	k	0	0	0	0
k	0	0	0	0	0
l	0	0	0	0	k
m	$-l$	0	0	$-k$	j

[2312^3]. The defining equations of this case are

$$ml = lm = 0, \quad m^2 = c_5 k.$$

There are two cases:

[$2312^3 1$], when d_{51} is not unity;
[2312^4], when d_{51} is unity.

[$2312^3 1$]. The defining equation of this case is

$$d_{51} \neq 1,$$

which gives

$$i[(1-d_{51})m - c_{51}j] = (1-d_{51})l - c_{51}k, \quad i[(1-d_{51})l - c_{51}k] = 0,$$
$$[(1-d_{51})l - c_{51}k]i = 0, \quad [(1-d_{51})m - c_{51}j]i = d_{51}[(1-d_{51})l - c_{51}k],$$
$$[(1-d_{51})l - c_{51}k][(1-d_{51})m - c_{51}j] = 0,$$
$$[(1-d_{51})m - c_{51}j][(1-d_{51})l - c_{51}k] = 0,$$
$$[(1-d_{51})m - c_{51}j]^2 = (1-d_{51})^2 c_5 k;$$

so that the substitution of $(1-d_{51})m - c_{51}j$ for m, and of $(1-d_{51})l - c_{51}k$ for l, is the same as to make

$$c_{51} = 0.$$

There are now two cases:

[$2312^3 1^2$], when c_5 does not vanish;
[$2312^3 12$], when c_5 vanishes.

[$2312^3 1^2$]. The defining equation of this case can be reduced to

$$m^2 = k.$$

This gives a quintuple algebra which may be called (al_5), its multiplication table being*

(al_5)	i	j	k	l	m
i	j	k	0	0	l
j	k	0	0	0	0
k	0	0	0	0	0
l	0	0	0	0	0
m	dl	0	0	0	k

[$2312^3 12$]. The defining equation of this case is

$$m^2 = 0.$$

This gives a quintuple algebra which may be called (am_5), its multiplication table being

(am_5)	i	j	k	l	m
i	j	k	0	0	l
j	k	0	0	0	0
k	0	0	0	0	0
l	0	0	0	0	0
m	dl	0	0	0	0

---

* In relative form, $i = A:B + B:C + C:D + dE:F$, $j = A:C + B:D$, $k = A:D$, $l = A:F$, $m = A:E + B:F + E:D$. [C. S. P.]

[2312⁴]. The defining equation of this case is

$$d_{51} = 1.$$

There are two cases:

[2312⁴1], when c_{51} does not vanish;
[2312⁵], when c_{51} vanishes.

[2312⁴1]. The defining equation of this case is easily reduced to

$$c_{51} = 1.$$

There are two cases:

[2312⁴1²], when c_5 does not vanish;
[2312⁴12], when c_5 vanishes.

[2312⁴1²]. The defining equation of this case is easily reduced to

$$m^2 = k.$$

This gives a quintuple algebra which may be called (an_5), its multiplication table being *

(an_5)	i	j	k	l	m
i	j	k	0	0	l
j	k	0	0	0	0
k	0	0	0	0	0
l	0	0	0	0	0
m	$l+k$	0	0	0	k

[2312⁴12]. The defining equation of this case is

$$m^2 = 0.$$

This gives a quintuple algebra which may be called (ao_5), its multiplication table being †

---

* In relative form, $i = A:E + A:B + B:C + C:D + E:F$, $j = A:C + B:D + A:F$, $k = A:D$, $l = A:F$, $m = A:C + A:E + E:D$. [C. S. P.]

† In relative form, $i = A:B + B:C + C:D + E:F$, $j = A:C + B:D$, $k = A:D$, $l = A:F$, $m = A:C + A:E + B:F$. [C. S. P.]

(ao_5)	i	j	k	l	m
i	j	k	0	0	l
j	k	0	0	0	0
k	0	0	0	0	0
l	0	0	0	0	0
m	$l+k$	0	0	0	0

[2312^5]. The defining equation of this case is

$$mi = l.$$

There are two cases :

[$2312^5 2$], when c_5 does not vanish ;
[2312^6],   when c_5 vanishes.

[$2312^5 1$]. The defining equation of this case can be reduced to

$$m^2 = k.$$

This gives a quintuple algebra which may be called (ap_5), its multiplication table being*

(ap_5)	i	j	k	l	m
i	j	k	0	0	l
j	k	0	0	0	0
k	0	0	0	0	0
l	0	0	0	0	0
m	l	0	0	0	k

---

*In relative form, $i = A:B + B:C + C:D + E:F$; $j = A:C + B:D$, $k = A:D$, $l = A:F$, $m = A:E + B:F + E:D$.   [C. S. P.]

[2312^6]. The defining equation of this case is
$$m^2 = 0.$$

This gives a quintuple algebra which may be called (aq_5), its multiplication table being

(aq_5)	i	j	k	l	m
i	j	k	0	0	l
j	k	0	0	0	0
k	0	0	0	0	0
l	0	0	0	0	0
m	l	0	0	0	0

[232]. The defining equation of this case is
$$im = 0,*$$

which gives
$$0 = jm = km,$$

and it may be assumed that
$$li = 0.$$

This gives
$$lj = lk = 0 = il^2 = l^2i = ilm = iml = mli = im.$$

There are two cases:

$$[2321], \text{ when } mi = l;$$
$$[232^2], \text{ when } mi = 0.$$

[2321]. The defining equation of this case is
$$mi = l,$$

which gives
$$0 = mj = mk, \quad lm = c_{45}k + d_{45}l + e_{45}m,$$
$$lmi = l^2 = e_{45}l, \quad 0 = l^4 = e_{45}l^3 = e_{45} = l^2, \quad m^2 = c_5k + d_5l + e_5m,$$
$$m^2i = ml = e_5l, \quad 0 = m^4l = e_5m^3l = e_5 = ml; \quad 0 = lm^2 = d_{45}lm = d_{45}.$$

---

* What is meant is that every quantity not involving powers of i is nilfaciend with reference to i. Hence, $il = 0$, also. [C. S. P.]

There are two cases:

$[2321^2]$, when c_{45} does not vanish;

$[23212]$, when c_{45} vanishes.

$[2321^2]$. The defining equation of this case can be reduced to

$$lm = k,^*$$

which gives

$$m^2 = c_5 k, \quad (m - c_5 l)^2 = 0,$$

so that the substitution of $m - c_5 l$ for m is the same as to make

$$c_5 = 0.$$

This gives a quintuple algebra which may be called (ar_5), of which the multiplication table is

(ar_5)	i	j	k	l	m
i	j	k	0	0	0
j	k	0	0	0	0
k	0	0	0	0	0
l	0	0	0	0	k
m	l	0	0	0	0

$[23212]$. The defining equation of this case is

$$lm = 0.$$

There are two cases:

$[232121]$, when d_5 does not vanish;

$[23212^2]$, when d_5 vanishes.

$[232121]$. The defining equation of this case can be reduced to

$$d_5 = 1.$$

There are two cases:

$[232121^2]$, when c_5 does not vanish;

$[2321212]$, when c_5 vanishes.

$[232121^2]$. The defining equation of this case can be reduced to

$$c_5 = 1.$$

---

* But $0 = im = mim = lm$. Thus, this case disappears, and the algebra (ar_5) is incorrect. [C. S. P.]

This gives a quintuple algebra which can be called (as_5), its multiplication table being*

(as_5)	i	j	k	l	m
i	j	k	0	0	0
j	k	0	0	0	0
k	0	0	0	0	0
l	0	0	0	0	0
m	l	0	0	0	$k+l$

[2321212]. The defining equation of this case is

$$c_5 = 0.$$

This gives a quintuple algebra which may be called (at_5), its multiplication table being

(at_5)	i	j	k	l	m
i	j	k	0	0	0
j	k	0	0	0	0
k	0	0	0	0	0
l	0	0	0	0	0
m	l	0	0	0	l

[23212^2]. The defining equation of this case is

$$m^2 = c_5 k.$$

There are two cases:

[23212^{2}1], when c_5 does not vanish;
[23212^3], when c_5 vanishes.

---

*In relative form, $i = A:B + B:C + C:D + E:F$, $j = A:C + B:D$, $k = A:D$, $l = A:F$, $m = A:E + E:F + E:D$. Omitting the last term of m, we have (at_5). [C. S. P.]

[23212²1]. The defining equation of this case can be reduced to

$$m^2 = k.$$

This gives a quintuple algebra which may be called (au_5), its multiplication table being *

(au_5)	i	j	k	l	m
i	j	k	0	0	0
j	k	0	0	0	0
k	0	0	0	0	0
l	0	0	0	0	0
m	l	0	0	0	k

[23212³]. The defining equation of this case is

$$m^2 = 0.$$

This gives a quintuple algebra which may be called (av_5), its multiplication table being

(av_5)	i	j	k	l	m
i	j	k	0	0	0
j	k	0	0	0	0
k	0	0	0	0	0
l	0	0	0	0	0
m	l	0	0	0	0

---

* In relative form, $i = A:B + B:C + C:D$, $j = A:C + B:D$, $k = A:D$, $l = E:D$, $m = E:C + A:F + F:D$. The omission of the last two terms of m gives (av_5). [C. S. P.]

[232²]. The defining equation of this case is

$$mi = 0,$$

which gives

$$0 = mj = mk = lmi = m^2 i,$$

and there is no pure algebra in this case.

[24]. The defining equation of this case is

$$i^3 = 0,$$

and by § 59,

$$i^2 = j, \quad ij = ji = j^2 = 0.$$

There are three cases:

[241], when $ik = l$, $il = m$;
[242], when $ik = l$, $il = im = 0$;
[243], when $ik = il = im = 0$.

[241]. The defining equations of this case are

$$ik = l, \quad il = m,$$

which give

$jk = m, \quad im = jl = jm = 0, \quad 0 = iml = ml^3 = e_{54}ml^2 = e_{54}, \quad jk = m,$

$il^2 = ml = b_{54}j, \quad l^2 = b_{54}i + b_4j + e_4m, \quad 0 = l^3 = b_{54}m + e_4ml = b_{54} = ml,$

$im^2 = 0, \quad m^2 = b_5j + e_5m, \quad 0 = m^3 = e_5m^2 = e_5,$

$imi = 0, \quad mi = b_{51}j + e_{51}m, \quad mj = e_{51}mi, \quad mi^3 = 0 = e_{51},$

$ili = mi = b_{51}j, \quad li = b_{51}i + b_{41}j + e_{41}m,$

$lil = lm = b_{51}m, \quad 0 = l^3m = b_{51} = lm = mi = mil = m^2, \quad (li)i = lj,$

$ik^2 = llk = a_3j + c_3l + d_3m, \quad illk = mk = c_3m, \quad lik = l^2 = a_{31}m,$

$0 = mk^3 = c_3^3m = c_3 = mk = k^2m,$

$kj = ki^2 = a_{31}j + d_{31}li = a_{31}(1 + d_{31})j + d_{31}^2m,$

$kil = km = a_{31}(1 + d_{31})m, \quad 0 = k^2m = a_{31}(1 + d_{31})km = a_{31}(1 + d_{31}) = km,$

$kj = d_{31}^2m, \quad 0 = k^3 = a_3l + b_3m + d_3lk = a_3 = b_3 + d_3^2 = b_3kj + d_3kl,$

$kl = a_{31}l + (b_{31} + d_3d_{31})m, \quad 0 = klk = a_{31}lk = d_3a_{31} = lkl = a_{31}l^2 = a_{31} = l^2,$

$0 = d_3a_{31} = b_{31}d_3 + d_3^2d_{31} + b_3d_{31}^2,$

$*0 = ki^2 + iki + i^2k = (d_{31}^2 + d_{31} + 1)m, \quad d_{31} = \sqrt[3]{1} = \mathfrak{r},$

$0 = k^2i + kik + ik^2 = b_{31} + d_0(1 + 2d_{31}), \quad i(k + pi) = l + pj, \quad i(l + pj) = m,$

$(k + pi)i = b_{31}j + d_{31}l + e_{31}m + pj = (b_{31} + p - pd_{31})j + d_{31}(l + pj) + e_{31}m,$

$(l + pj)i = d_{31}m,$

---

* This line and the first equation of the next can be derived from $0 = (i + \mathfrak{J}k)^3$. [C. S. P.]

so that if p satisfies the equation

$$p\,(d_{31} - 1) = b_{31},$$

the substitution of $k + pi$ for k and of $l + pj$ for l is the same as to make

$$0 = b_{31} = d_3 = b_3.$$

There are four cases :

[241²], when neither e_{31} nor e_3 vanishes ;
[2412], when e_{31} does not vanish but e_3 vanishes ;
[2413], when e_{31} vanishes and not e_3 ;
[2414], when e_{31} and e_3 both vanish.

[241²]. The defining equations of this case can be reduced, without loss of generality, to

$$e_{31} = e_3 = 1.$$

We thus obtain a quintuple algebra which may be called (aw_5), its multiplication table being *

(aw_5)	i	j	k	l	m
i	j	0	l	m	0
j	0	0	m	0	0
k	$\mathfrak{r}l + m$	$\mathfrak{r}^2 m$	m	0	0
l	$\mathfrak{r}m$	0	0	0	0
m	0	0	0	0	0

[2412]. The defining equations of this case can be reduced to

$$e_{31} = 1, \quad e_3 = 0.$$

---

* In relative form $i = A:B + B:D + \mathfrak{r}C:E + \mathfrak{r}E:F + G:F$, $j = A:D + \mathfrak{r}^2C:F$, $k = A:C + B:E + D:F + A:G + G:F$, $l = A:E + B:F$, $m = A:F$. To obtain (ax_5), omit the last term of k. To obtain (ay_5), omit, instead, the last term of i. To obtain (az_5), omit both these last terms. [C. S. P.]

We thus obtain a quintuple algebra which may be called (ax_5), its multiplication table being

(ax_5)	i	j	k	l	m
i	j	0	l	m	0
j	0	0	m	0	0
k	$\mathfrak{r}l+m$	$\mathfrak{r}^2m$	0	0	0
l	$\mathfrak{r}m$	0	0	0	0
m	0	0	0	0	0

[2413]. The defining equations of this case can be reduced to

$$e_{31} = 0, \quad e_3 = 1.$$

We thus obtain a quintuple algebra which may be called (ay_5), its multiplication table being

(ay_5)	i	j	k	l	m
i	j	0	l	m	0
j	0	0	m	0	0
k	$\mathfrak{r}l$	$\mathfrak{r}^2m$	m	0	0
l	$\mathfrak{r}m$	0	0	0	0
m	0	0	0	0	0

[2414]. The defining equations of this case are

$$e_{31} = e_3 = 0.$$

We thus obtain a quintuple algebra which may be called (az_5), its multiplication table being

(az_5)	i	j	k	l	m
i	j	0	l	m	0
j	0	0	m	0	0
k	$\mathfrak{r}l$	$\mathfrak{r}^2m$	0	0	0
l	$\mathfrak{r}m$	0	0	0	0
m	0	0	0	0	0

[242]. The defining equations of this case are

$$ik = l, \quad il = im = 0,$$

which give

$$0 = jk = jl = jm,$$
$$li = iki = a_{31}j + c_{31}l, \quad 0 = li^3 = c_{31} = li^2 = lj,$$
$$0 = a_{31}jk = lik = l^2 = ikl = a_{34} = c_{34},$$
$$ik^2 = lk = a_3j + c_3l, \quad 0 = ik^3 = lk^2 = c_3lk = c_3,$$
$$0 = imi = a_{51} = c_{51}, \quad mi^2 = mj = d_{51}li + e_{51}mi, \quad 0 = mji = e_{51},$$
$$mj = a_{31}d_{51}j, \quad 0 = m^3j = a_{31}d_{51} = mj,$$
$$ikm = lm = a_{35}j + c_{35}l, \quad 0 = lm^3 = c_{35},$$
$$0 = i^2k + iki + ki^2 = e_{31}d_{51} = 2a_{31} + a_{31}d_{31} + e_{31}b_{51},$$
$$kj = -a_{31}j, \quad 0 = k^3j = a_{31} = kj = li = e_{31}b_{51},$$
$$0 = imk = a_{53} = c_{53}, \quad 0 = mk^3 = e_{53},$$
$$mik = ml = a_3d_{51}j, \quad kik = kl = (a_3d_{31} + e_{31}b_{53})j + e_{31}d_{53}l, \quad e_{34} = 0,$$
$$0 = k^3l = e_{31}d_{53} = lki = e_{31}lm = e_{31}a_{35} = k^3m = e_{35}.$$

There are two cases :

[2421], when e_{31} does not vanish ;
[242²], when e_{31} vanishes.

[2421]. The defining equation of this case can be reduced to

$$ki = m,$$

which, by the aid of the above equations, gives

$$0 = mi = kil = ml = kim = m^2, \quad a_3 j = ik^2 = k^2 i = lk = km, \quad 0 = lm,$$
$$b_{53} j = kik = kl = mk, \quad 0 = ik^2 + kik + k^2 i = 2a_3 + b_{53},$$
$$0 = k^3 = a_3 = b_{53} = kl = km = mk = ml;$$

and if p is determined by the equation

$$p^2 + (d_3 + e_3)\,p - b_3 = 0,$$

$k + pi$, $l + pj$, and $m + pj$ can be respectively substituted for k, l and m, which is the same thing as to make

$$b_3 = 0.$$

There are three cases:

$[2421^2]$, when neither d_3 nor e_3 vanishes;
$[24212]$, when d_3 vanishes and not e_3;
$[24213]$, when d_3 and e_3 both vanish.

$[2421^2]$. The defining equation of this case can be reduced to

$$d_3 = 1.$$

This gives a quintuple algebra which may be called (ba_5), its multiplication table being *

(ba_5)	i	j	k	l	m
i	j	0	l	0	0
j	0	0	0	0	0
k	m	0	$l+em$	0	0
l	0	0	0	0	0
m	0	0	0	0	0

---

* In relative form, $i = A:B + B:C + A:E$, $j = A:C$, $k = D:B + E:F + D:G + eG:C + A:E$, $l = A:F$, $m = D:C$. By omitting the last term of k and putting $e = 1$ we get (bb_5), and by omitting the last two terms of k we get (bc_5). [C. S. P.]

[2421²]. The defining equation of this case can be reduced to

$$k^2 = m.$$

This gives a quintuple algebra which may be called (bb_5), its multiplication table being

(bb_5)	i	j	k	l	m
i	j	0	l	0	0
j	0	0	0	0	0
k	m	0	m	0	0
l	0	0	0	0	0
m	0	0	0	0	0

[24213]. The defining equation of this case is

$$k^2 = 0.$$

This gives a quintuple algebra which may be called (bc_5), its multiplication table being

(bc_5)	i	j	k	l	m
i	j	0	l	0	0
j	0	0	0	0	0
k	m	0	0	0	0
l	0	0	0	0	0
m	0	0	0	0	0

[242²]. The defining equation of this case is

$$e_{31} = 0 .$$

There are two cases :

[242²1], when e_3 does not vanish ;
[242³], when e_3 vanishes.

[242²1]. The defining equation of this case can be reduced to

$$k^2 = a_3 i + m ,$$

which gives

$kik = kl = a_3 d_{31} j , \quad ik^2 = lk = a_3 j , \quad k^2 i = a_3 j + mi = d_{31} kl = a_3 d_{31}^2 j ,$
$\quad 0 = k^2 i + ik + kik = a_3 (d_{31}^2 + d_{31} + 1) , \quad mi = a_3 (d_{31} l - 1) j , \quad ml = mik = 0 ,$
$\quad 0 = k^3 = a_3 l + mk = a_3 ki + km ,$
$mk = - a_3 l , \quad km = - a_3 b_{31} j - a_3 d_{31} l , \quad lm = 0 .$

There are two cases :

[242²1²], when a_3 does not vanish ;
[242²12], when a_3 vanishes.

[242²1²]. The defining equation of this case can be reduced to

$$k^2 = i + m ,$$

which gives

$$d_{31} = \sqrt[3]{1} = \mathfrak{r} , \quad lk = j , \quad mk = - l ,$$
$$ki = - km = b_{31} j + \mathfrak{r} l , \quad mi = (\mathfrak{r}^2 - 1) j , \quad m^2 = - \mathfrak{r} j .$$

There are two cases :

[242²1³], when b_{31} does not vanish ;
[242²1²2], when b_{31} vanishes.

[242²1³]. The defining equation of this case can be reduced to

$$ki = j + \mathfrak{r} l .$$

This gives a quintuple algebra which may be called (bd_5), its multiplication table being *

---

* In relative form, $i = A : D + D : F + B : E + C : F , \; j = A : F , \; k = \mathfrak{r} A : B + \mathfrak{r} B : C + D : E - \frac{1}{\mathfrak{r}} D : F$
$+ E : F , \; i = A : E - \frac{1}{\mathfrak{r}} A : F + B : F , \; m = \mathfrak{r}^2 A : C - A : D - B : E - C : F .$ [C. S. P.]

(bd_5)	i	j	k	l	m
i	j	0	l	0	0
j	0	0	0	0	0
k	$j+\mathfrak{r}l$	0	$i+m$	$\mathfrak{r}j$	$-j-\mathfrak{r}l$
l	0	0	j	0	0
m	$(\mathfrak{r}^2-1)j$	0	$-l$	0	$-\mathfrak{r}^2j$

$[242^2 1^2 2]$. The defining equation of this case is

$$ki = \mathfrak{r}l .$$

This gives a quintuple algebra which may be called (be_5), its multiplication table being *

(be_5)	i	j	k	l	m
i	j	0	l	0	0
j	0	0	0	0	0
k	$\mathfrak{r}l$	0	$i+m$	$\mathfrak{r}j$	$-\mathfrak{r}l$
l	0	0	j	0	0
m	$(\mathfrak{r}^2-1)j$	0	$-l$	0	$-\mathfrak{r}^2j$

$[242^2 12]$. The defining equation of this case is

$$k^2 = m ,$$

which gives

$$0 = kl = lk = km = mk = m^2 = k^2i = mi .$$

There are two cases :

$[242^2 121]$, when b_{31} does not vanish;

$[242^2 12^2]$, when b_{31} vanishes.

---

* On adding to the expression for k in the last note the term $- A : C$, we have this algebra in relative form. [C. S. P.]

[242^2121]. The defining equation of this case can be reduced to

$$ki = j + d_{31}l.$$

This gives a quintuple algebra which may be called (bf_5), its multiplication table being *

(bf_5)	i	j	k	l	m
i	j	0	l	0	0
j	0	0	0	0	0
k	$j+dl$	0	m	0	0
l	0	0	0	0	0
m	0	0	0	0	0

[242^212^2]. The defining equation of this case is

$$ki = d_{31}l.$$

This gives a quintuple algebra which may be called (bg_5), its multiplication table being †

(bg_5)	i	j	k	l	m
i	j	0	l	0	0
j	0	0	0	0	0
k	dl	0	m	0	0
l	0	0	0	0	0
m	0	0	0	0	0

---

* In relative form, $i = A : B + B : C + D : E$, $j = A : C$, $k = A : B + dA : D + B : E + B : F$, $l = A : E$, $m = A : E + A : F$. [C. S. P.]

† In relative form, $i = A : B + B : C + D : E$, $j = A : C$, $k = dA : D + B : E + B : F$, $l = A : E$, $m = A : F$. The algebra (ar_5) is what this becomes when $d = 0$. [C. S. P.]

[242³]. The defining equation of this case is

$$e_3 = 0,$$

which gives *

---

* It is not easy to see how the author proves that $a_3 = 0$. But it can be proved thus. $0 = k^3 = (a_3 i + b_3 j + d_3 l) k = a_3 l + a_3 d_3 j$.

The algebras of the case [242³] are those quintuple systems in which every product containing j or l as a factor vanishes, while every product which does not vanish is a linear function of j and l. Any multiplication table conforming to these conditions is self-consistent, but it is a matter of some trouble to exclude every case of a *mixed* algebra. An algebra of the class in question is separable, if all products are similar. But this case requires no special attention ; and the only other is when two dissimilar expressions U and V can be found, such that both being linear functions of i, k and m, $UV = VU = 0$. It will be convenient to consider separately, first, the conditions under which $UV - VU = 0$, and, secondly, those under which $UV + VU = 0$. To bring the subjects under a familiar form, we may conceive of i, k, m as three vectors not coplanar, so that, writing

$$U = xi + yk + zm, \qquad V = x'i + y'k + z'm,$$

we have x, y, z, and x', y', z', the Cartesian coördinates of two points in space. [We might imagine the space to be of the hyperbolic kind, and take the coëfficients of j and l as coördinates of a point on the quadric surface at infinity. But this would not further the purpose with which we now introduce geometric conceptions.] But since we are to consider only such properties of U and V as belong equally to all their numerical multiples, we may assume that they always lie in any plane

$$Ax + By + Cz = 1,$$

not passing through the origin ; and then x, y, z, and x', y', z', will be the homogeneous coördinates of the two points U and V in that plane. Let it be remembered that, although i, k, m are vectors, yet their multiplication does not all follow the rule of quaternions, but that

$$i^2 = b_1 j + d_1 l, \qquad ik = b_{13} j + d_{13} l, \qquad im = b_{15} j + d_{15} l,$$
$$ki = b_{31} j + d_{31} l, \qquad k^2 = b_3 j + d_3 l, \qquad km = b_{35} j + d_{35} l,$$
$$mi = b_{51} j + d_{51} l, \qquad mk = b_{53} j + d_{53} l, \qquad m^2 = b_5 j + d_5 l.$$

The condition that $UV - VU = 0$ is expressed by the equations

$$(b_{13} - b_{31})(xy' - x'y) + (b_{15} - b_{51})(xz' - x'z) + (b_{35} - b_{53})(yz' - y'z) = 0,$$
$$(d_{13} - d_{31})(xy' - x'y) + (d_{15} - d_{51})(xz' - x'z) + (d_{35} - d_{53})(yz' - y'z) = 0.$$

The first equation evidently signifies that for every value of U, V must be on a straight line, that this line passes through U, and that it also passes through the point

$$P = (b_{35} - b_{53}) i + (b_{51} - b_{15}) k + (b_{13} - b_{31}) m.$$

The second equation expresses that the line between U and V contains the point

$$Q = (d_{35} - d_{53}) i + (d_{51} - d_{15}) k + (d_{13} - d_{31}) m.$$

The two equations together signify, therefore, that U and V may be any two points on the line between the fixed points P and Q. Linear transformations of j and l may shift P and Q to any other situations on the line joining them, but cannot turn the line nor bring the two points into coincidence.

The condition that $UV + VU = 0$ is expressed by the equations

$$2b_1 xx' + 2b_3 yy' + 2b_5 zz' + (b_{13} + b_{31})(xy' + x'y) + (b_{15} + b_{51})(xz' + x'z) + (b_{35} + b_{53})(yz' + y'z) = 0,$$
$$2d_1 xx' + 2d_3 yy' + 2d_5 zz' + (d_{13} + d_{31})(xy' + x'y) + (d_{15} + d_{51})(xz' + x'z) + (d_{35} + d_{53})(yz' + y'z) = 0.$$

The first of these evidently signifies that for any position of V the locus of U is a line ; that U being fixed at any point on that line, V may be carried to any position on a line passing through its original position ; and that further, if U is at one of the two points where its line cuts the conic

$$b_1 x^2 + b_3 y^2 + b_5 z^2 + (b_{13} + b_{31}) xy + (b_{15} + b_{51}) xz + (b_{35} + b_{53}) yz = 0,$$

$$0 = k^2 i = a_3 j = a_3 = lk = ml = kl = m^3 = e_5 = d_5 a_{35} = k^3 m = e_{35},$$
$$0 = kmk = a_{35} l = a_{35} = lm.$$

then V may be at an infinitely neighboring point on the same conic, so that tangents to the conic from V cut the locus of U at their points of tangency. The second equation shows that the points U and V have the same relation to the conic

$$d_1 x^2 + d_3 y^2 + d_5 z^2 + (d_{13} + d_{31}) xy + (d_{15} + d_{51}) xz + (d_{35} + d_{53}) yz = 0.$$

These conics are the loci of points whose squares contain respectively no term in j and no term in l. Their four intersections represent expressions whose squares vanish. Hence, linear transformations of j and l will change these conics to any others of the sheaf passing through these four fixed points. The two equations together, then, signify that through the four fixed points, two conics can be drawn tangent at U and V to the line joining these last points.

Uniting the conditions of $UV - VU = 0$ and $UV + VU = 0$, they signify that U and V are on the line joining P and Q at those points at which this line is tangent to conics through the four fixed points whose squares vanish. But if the algebra is pure, it is impossible to find two such points; so that the line between P and Q must pass through one of the four fixed points. In other words, the necessary condition of the algebra being pure is that one and only one nilpotent expression in i, k, m, should be a linear function of P and Q.

The two points P and Q together with the two conics completely determine all the constants of the multiplication table. Let S and T be the points at which the two conics separately intersect the line between P and Q. A linear transformation of j will move P to the point $pP + (1-p) Q$ and will move S to the point $pS + (1-p) T$, and a linear transformation of l will move Q and T in a similar way. The points P and S may thus be brought into coincidence, and the point Q may be brought to the common point of intersection of the two conics with the line from P to Q. The geometrical figure determining the algebra is thus reduced to a first and a second conic and a straight line having one common intersection. This figure will have special varieties due to the coincidence of different intersections, etc.

There are six cases : [1], there is a line of quantities whose squares vanish and one quantity out of the line ; [2], there are four dissimilar quantities whose squares vanish ; [3], two of these four quantities coincide ; [4], two pairs of the four quantities coincide ; [5], three of the four quantities coincide ; [6], all the quantities coincide.

We may, in every case, suppose the equation of the plane to be $x + y + z = 1$.

[1]. In this case, the line common to the two conics may be taken as $y = 0$, and the separate lines of the conics as $z = 0$ and $x = 0$, respectively. We may also assume $2P = x + y$ and $2Q = x + z$. We thus obtain the following multiplication table, where the rows and columns having j and l as their arguments are omitted :

	i	k	m
i	0	$3l$	$-j$
k	$-l$	0	$3j+l$
m	j	$l-j$	0

[2]. In this case, we may take k as the common intersection of the two conics and the line, i, m, and $i - k + m$ as the other intersections of the conics. We have $Q = k$, and we may write

$$P = S = pi + (1-p-q)k + qm, \qquad T = rP + (1-r)Q = rpi + (1-rp-rq)k + rqm.$$

We thus obtain the following multiplication table :

There are two cases :

$$[242^31], \text{ when } d_5 \text{ does not vanish};$$
$$[242^4], \quad \text{when } d_5 \text{ vanishes}.$$

$[242^31]$. The defining equation of this case can be reduced to

$$d_5 = 1,$$

which gives

and if

$$i\,(k + b_5 i + pm) = l + b_5 j = m^2\;;$$

$$p^2 = -\,d_3 - b_5 - b_5 d_{31} - b_5 p d_{51} - p d_{35} - p d_{53},$$

	i	k	m
i	0	$q\,(q+1)j + rq\,(rq-1)\,l$	$[-2-p\,(p-3)+q\,(q+1)]j$ $+[-2-rp(rp-1)+rq\,(rq-1)]l$
k	$q\,(q-3)j + rq\,(rq-1)\,l$	0	$-p\,(p-3)j - rp\,(rp-1)\,l$
m	$[2-p\,(p+1)+q\,(q-3)]j+$ $[2-rp\,(rp-1)+rq\,(rq-1)]l$	$-p\,(p+1)j - rp\,(rp-1)\,l$	0

[3]. Let k be the double point common to the two conics, and let i and m be their other intersections. Then all expressions of the form $ku + uk$ are similar. The line between P and Q cannot pass through k, because in that case all products would be similar. We may therefore assume that it passes through i. Then, we have $Q = i$, we may assume $S = P = i - k + m$, and we may write $T = rP + (1-r)\,Q = i - rk + rm$. The equation of the common tangent to the conics at k may be written $hx + (1-h)z = 0$. Then the equations of the two conics are

$$hxy + xz + (1-h)yz = 0,$$
$$hxy + (h+r-hr)xz + (1-h)yz = 0.$$

We thus obtain the following multiplication table :

	i	k	m
i	0	$(h+1)j + (h+r)\,l$	$2j + [h\,(1-r)+2r]\,l$
k	$(h-1)j + (h-r)\,l$	0	$(2-h)(j+l)$
m	$h\,(1-r)\,l$	$-h\,(j+l)$	0

[4]. In this case we may take i and m as the two points of contact of the conics, k as P, and $i-k+m$ as T. Then writing the equations of the two tangents

$$gy + z = 0. \qquad x + hy = 0.$$

the two conics become

$$gxy + xz + hyz = 0,$$
$$(g+h-1)\,y^2 + gxy + xz + hyz = 0,$$

and the multiplication table is as follows :

the substitution of $k + b_5 i + pm$ for k and $l + b_5 j$ for l is the same as to make

$$b_5 = d_3 = 0 .$$

This gives

$$m^2 = l , \quad k^2 = b_3 j .$$

There are two cases:

$$[242^3 1^2], \text{ when } b_3 \text{ does not vanish;}$$
$$[242^3 12], \text{ when } b_3 \text{ vanishes.}$$

$[242^3 1^2]$. The defining equation of this case can be reduced to

$$k^2 = j .$$

	i	k	m
i	$(g+h-1)l$	$gj+(g+1)l$	$2l$
k	$gj+(g-1)l$	0	$hj+(h+1)l$
m	$2j$	$hj+(h-1)l$	0

[5]. In this case, we may take k as the point of osculation of the conics and i as their point of intersection. The line between P and Q must either, [51], pass through k, or, [52], pass through i.

[51]. We may, without loss of generality, take

$$P = k , \quad Q = m ,$$

and the equations of the two conics are

$$z^2 + rxz = 0 , \quad rxy + 2qxz + 2yz = 0 .$$

Then, the multiplication table is as follows:

	i	k	m
i	0	0	ql
k	rl	0	l
m	$rj+ql$	l	j

[52]. We have $Q = i$, we may take $T = m$, and we may assume $P = 2i - m$ and $b_{13} + b_{31} = 1$. Then, we may write the equations of the two conics,

$$2z^2 + xy + xz + ryz = 0 ,$$
$$- rxy + (2-r) xz + r^2 yz = 0 .$$

We thus obtain the following multiplication table:

This gives a quintuple algebra which can be called (bh_5), its multiplication table being *

(bh_5)	i	j	k	l	m
i	j	0	l	0	0
j	0	0	0	0	0
k	$aj+bl$	0	j	0	$cj+dl$
l	0	0	0	0	0
m	$a'j+b'l$	0	$c'j+d'l$	0	$l \cdot$

	i	k	m
i	0	$-rl$	$j-(r-2)l$
k	$2j-rl$	0	$(r-2)j+(r^2+1)l$
m	$j-(r-2)l$	$(r-2)j+(r^2-1)l$	$2j$

[6]. The conics have but one point in common. This may be taken at k. We have $Q=k$, we may take $T=i$ and $2P=2S=i+k$. We may also take $b_1=-1$. Then the equations of the two conics may be written

$$-x^2+pz^2+2xy+4qxz+2ryz=0,$$
$$(4+pr^2)z^2+2xy+4(q+r)xz+2ryz=0.$$

We thus find this multiplication table:

	i	k	m
i	$-j$	$j+l$	$(2q-1)j+2(q+r-p)l$
k	$j+l$	0	$(r+1)j+rl$
m	$(2q+1)+2(q+r+p)l$	$(r-1)j+rl$	$pj+(4+pr^2)l$

If this analysis is correct, only three indeterminate coefficients are required for the multiplication tables of this class of algebras. [C. S. P.]

* See last note. I do not give relative forms for this class of algebras, owing to the extreme ease with which they may be found. [C. S. P.]

[$242^3 12$]. The defining equation of this case is

$$k^2 = 0.$$

There are two cases:

[$242^3 121$], when b_{31} does not vanish;
[$242^3 12^2$], when b_{31} vanishes.

[$242^3 121$]. The defining equation of this case can be reduced to

$$b_{31} = 1.$$

This gives a quintuple algebra which may be called (bi_5), its multiplication table being

(bi_5)	i	j	k	l	m
i	j	0	l	0	0
j	0	0	0	0	0
k	$j+al$	0	0	0	$bj+cl$
l	0	0	0	0	0
m	$a'j + b'l$	0	$c'j + d'l$	0	l

[$242^3 12^2$]. The defining equation of this case is

$$ki = d_{31}l.$$

There are two cases:

[$242^3 12^2 1$], when b_{51} does not vanish;
[$242^3 12^3$], when b_{51} vanishes.

[$242^3 12^2 1$]. The defining equation of this case can be reduced to

$$b_{51} = 1.$$

This gives a quintuple algebra which may be called (bj_5), its multiplication table being

(bj_5)	i	j	k	l	m
i	j	0	l	0	0
j	0	0	0	0	0
k	al	0	0	0	$bj+cl$
l	0	0	0	0	0
m	$j+a'l$	0	$b'j+il$	0	l

[$242^2 12^3$]. The defining equation of this case is

$$mi = d_{51}l;$$

which can always, in the case of a pure algebra, be reduced to

$$mi = l.$$

This gives a quintuple algebra which may be called (bk_5), its multiplication table being

(bk_5)	i	j	k	l	m
i	j	0	l	0	0
j	0	0	0	0	0
k	al	0	0	0	$bj+cl$
l	0	0	0	0	0
m	l	0	$a'j+b'l$	0	l

[242^4]. The defining equation of this case is

$$m^2 = b_5 j,$$

and it can be reduced to $[242^31]$ unless

$$d_{51} = d_3 = 0, \quad k^2 = b_3j, \quad d_{31} = -1, \quad d_{35} = -d_{53};$$

whence it may be assumed that

$$mi = j;$$

and since

$$(k + bi)^2 = 0,$$

when

$$p^2 + pb_{31} + b_3 = 0,$$

it may also be assumed that

$$k^2 = 0.$$

There are two cases:

$[242^41]$, when b_{31} does not vanish;
$[242^5]$, when b_{31} vanishes.

$[242^41]$. The defining equation of this case can be reduced to

$$b_{31} = 1.$$

This gives a quintuple algebra which may be called (bl_5), its multiplication table being

(bl_5)	i	j	k	l	m
i	j	0	l	0	0
j	0	0	0	0	0
k	$j-l$	0	0	0	$aj+bl$
l	0	0	0	0	0
m	j	0	$aj+bl$	0	cj

$[242^5]$. The defining equation of this case is

$$ki = -l$$

There are two cases:

$[242^51]$, when b_{35} does not vanish;
$[242^6]$, when b_{35} vanishes.

[$242^5 1$]. The defining equation of this case can be reduced to

$$b_{35} = 1 \,.$$

This gives a quintuple algebra which may be called (bm_5), its multiplication table being *

(bm_5)	i	j	k	l	m
i	j	0	l	0	0
j	0	0	0	0	0
k	$-l$	0	0	0	$j+al$
l	0	0	0	0	0
m	j	0	$bj-al$	0	cj

[242^6]. The defining equation of this case is

$$b_{35} = 0 \,.$$

There are two cases :

> [$242^6 1$], when b_{53} does not vanish;
> [242^7], when b_{53} vanishes.

[$242^6 1$]. The defining equation of this case can be reduced to

$$b_{53} = 1 \,.$$

This gives a quintuple algebra which may be called (bn_5), its multiplication table being †

---

* This algebra is mixed. Namely, if $b \neq 1$, it separates on substituting $i_1 = (1 - b) i + k$, $k_1 = (1 - b) i + [a (1 - b) + 1] k - (1 - b)^2 m$; but if $b = 1$, it separates on substituting $i_1 = ai - (a^2 + a + c) k + m$, $k_1 = ai + qk + m$. [C. S. P.]

† Substitute $i_1 = i - k$, $k_1 = ak + m$, and the algebra separates. [C. S. P.]

(bn_5)	i	j	k	l	m
i	j	0	l	0	0
j	0	0	0	0	0
k	$-l$	0	0	0	al
l	0	0	0	0	0
m	j	0	$j-al$	0	cj

[242⁷]. The defining equation of this case is

$$b_{53} = 0.$$

This gives a quintuple algebra which may be called (bo_5), its multiplication table being *

(bo_5)	i	j	k	l	m
i	j	0	l	0	0
j	0	0	0	0	0
k	$-l$	0	0	0	al
l	0	0	0	0	0
m	j	0	$-al$	0	cj

[243]. The defining equations of this case are

$$0 = ik = il = im,$$

which give

$$0 = jk = jl = jm.$$

---

* Substitute for m, $ai + m$, and the algebra separates.   [C. S. P.]

There are two cases:

$$[2431], \text{ when } ki = l, \; li = m, \; mi = 0 \, ;$$
$$[2432], \text{ when } ki = l, \; li = mi = 0.$$

[2431]. The defining equations of this case are

$$ki = l, \quad li = m, \quad mi = 0,$$

which give

$$kj = m, \quad lj = mj = 0 = lk = mk = l^2 = lm = ml = m^2,$$
$$0 = ik^2 = ikl = ikm = a_3 = a_{34} = a_{35},$$
$$k^2 i = kl = c_3 l + d_3 m,$$
$$kli = km = c_3 m, \quad 0 = k^3 m = c_3 = km,$$
$$0 = k^3 = b_3 m + d_3^2 m = b_3 + d_3^2.$$

There are two cases:

$$[2431^2], \text{ when } e_3 \text{ does not vanish};$$
$$[24312], \text{ when } e_3 \text{ vanishes.}$$

[2431²]. The defining equation of this case can be reduced to

$$e_3 = 1.$$

This gives a quintuple algebra which may be called (bp_5), its multiplication table being *

---

* The structure of this algebra may be exhibited by putting $k_1 = i + a^{-1}j - a^{-1}k$, $l_1 = j - a^{-1}l$, $m_1 = -a^{-1}m$, when the multiplication table becomes

	i	j	k	l	m
i	j	0	j	0	0
j	0	0	0	0	0
k	l	m	0	0	0
l	m	0	m	0	0
m	0	0	0	0	0

In relative form, $i = B:C + C:D$, $j = B:D$, $k = A:B + C:D$, $l = A:C$, $m = A:D$. [C. S. P.]

(bp_5)	i	j	k	l	m
i	j	0	0	0	0
j	0	0	0	0	0
k	l	m	$-a^2j+al+m$	am	0
l	m	0	0	0	0
m	0	0	0	0	0

[24312]. The defining equation of this case is

$$e_3 = 0 .$$

This gives a quintuple algebra which may be called (bq_5), its multiplication table being*

(bq_5)	i	j	k	l	m
i	j	0	0	0	0
j	0	0	0	0	0
k	l	m	$-a^2j+al$	am	0
l	m	0	0	0	0
m	0	0	0	0	0

[2432]. The defining equations of this case are

$$ki = l, \quad li - mi = 0 ,$$

---

* On substituting $k_1 = i - a^{-1}k$, $l_1 = j - a^{-1}l$, $m_1 = a^{-1}m$, this algebra reduces to (bp_5), in the form given in the last note. [C. S. P.]

which give

$$0 = kj = lj = mj = lk = l^2 = lm = ik^2 = a_3 ,$$
$$kl = c_3 l , \quad 0 = k^3 l = c_3 = kl ,$$
$$0 = ikm = a_{35} = kmi = c_{35} = k^3 m = e_{35} = imk = a_{53} ,$$
$$ml = c_{53} l , \quad 0 = m^3 l = c_{53} = ml = mk^3 = e_{53} ;$$

and it may be assumed that

$$k^2 = m ,$$

which gives

$$0 = k^3 = km = mk = m^2 .$$

There is then a quintuple algebra which may be called (br_5), its multiplication being *

(br_5)	i	j	k	l	m
i	j	0	0	0	0
j	0	0	0	0	0
k	l	0	m	0	0
l	0	0	0	0	0
m	0	0	0	0	0

[25]. The defining equations of this case are

$$0 = i^2 = j^2 = k^2 = l^2 = m^2 = ij + ji = ik + ki = il + li = im + mi ,$$
$$0 = jk + kj = jl + lj = jm + mj = kl + lk = km + mk = lm + ml ;$$

and it may be assumed that

$$ij = k = -ji , \quad il = m = -li ;$$

---

* In relative form, $i = D : E + E : F$, $j = D : F$, $k = A : B + B : F + C : E$, $l = C : F$, $m = A : F$. [C. S. P.]

which gives

$$0 = ik = ki = jk = kj = im = mi = km = mk = lm = ml,$$
$$ijk = kl = b_{24}k + d_{24}m = -ilj = -mj = jm,$$
$$0 = j^2m = d_{24}jm = d_{24} = kl^2 = b_{24}kl = b_{24} = kl = lk = jm = mj,$$
$$0 = j^2l = a_{24}k = a_{24},$$
$$i(c_{24}j + e_{24}l) = c_{24}k + e_{24}m,$$
$$j(c_{24}j + e_{24}k) = e_{24}(c_{24}k + e_{24}m),$$
$$l(c_{24}j + e_{24}l) = -c_{24}(c_{24}k + e_{24}m);$$

so that it is easy to see that there is no pure algebra in this case.

SEXTUPLE ALGEBRA.

There are two cases:

[1], when there is an idempotent basis;
[2], when the algebra is nilpotent.

[1]. The defining equation of this case is

$$i^3 = i.$$

There are 19 cases:

[1²], when all the other units but i are in the first group;
[12], when j, k, l, m are in the first and n in the second group;
[13], when j, k and l are in the first and m and n in the second group;
[14], when j, k and l are in the first, m in the second and n in the third group;
[15], when j and k are in the first and l, m and n in the second group;
[16], when j and k are in the first, l and m in the second and n in the third group;
[17], when j and k are in the first, l in the second, m in the third, and n in the fourth group;
[18], when j is in the first, and k, l, m and n in the second group;
[19], when j is in the first, k, l and m in the second, and n in the third group;
[10'], when j is in the first, k and l in the second, and m and n in the third group;
[11'], when j is in the first, k and l in the second, m in the third and n in the fourth group;
[12'], when j is in the first, k in the second, l in the third and m and n in the fourth group;

[13'], when j, k, l, m and n are in the second group;

[14'], when j, k, l and m are in the second and m in the third group;

[15'], when j, k and l are in the second and m and n are in the third group;

[16'], when j, k and l are in the second, m in the third, and n in the fourth group;

[17'], when j and k are in the second, l and m in the third, and n in the fourth group;

[18'], when j and k are in the second, l in the third, and m and n in the fourth group;

[19'], when j is in the second, k in the third, and l, m and n in the fourth group.

[1²]; The defining equations of this case are

$$ij = ji = j, \quad ik = ki = k, \quad il = li = l, \quad im = mi = m, \quad in = ni = n,$$

and the 54 algebras of this case deduced from (q_5) to (br_5) may be called (a_6) to (bb_6).*

[12]. The defining equations of this case are

$$ij = ji = j, \quad ik = ki = k, \quad il = li = l, \quad im = mi = m, \quad in = n, \quad ni = 0,$$

which give

$$0 = jn = nj = kn = nk = ln = nl = mn = nm = n^2,$$

so that there is no pure algebra in this case.

[13]. The defining equations of this case are

$$ij = ji = j, \quad ik = ki = k, \quad il = li = l, \quad im = m, \quad in = n, \quad mi = ni = 0.$$

There are four cases, which correspond to relations between the units of the first group similar to those of the quadruple algebras (a_4), (b_4), (c_4) or (d_4).

[131]. The defining equations of this case are

$$j^2 = k, \quad jk = kj = l, \quad jl = k^2 = kl = lj = lk = l^2 = 0;$$

and, in the result, we obtain

$$jm = n, \quad jn = km = kn = lm = ln = 0.$$

---

* The multiplication tables of these algebras, formed from the nilpotent quintuple algebras, in the same manner in which the first class of quintuple algebras are formed from the nilpotent quadruple algebras, have been omitted. [C. S. P.]

This gives a sextuple algebra which may be called (bc_6), of which the multiplication table is *

(bc_6)	i	j	k	l	m	n
i	i	j	k	l	m	n
j	j	k	l	0	n	0
k	k	l	0	0	0	0
l	l	0	0	0	0	0
m	0	0	0	0	0	0
n	0	0	0	0	0	0

[132]. The defining equations of this case are

$$j^2 = k = l^2, \quad lj = ak, \quad jk = jl = kj = k^2 = kl = lk = 0,$$

which give

$$km = kn = 0.$$

There are two cases :

[132¹], when e_{24} does not vanish ;
[132²], when e_{24} vanishes.

[132¹]. The defining equation of this case can be reduced to

$$jn = m,$$

which gives

$$0 = jm = lm.$$

This gives a sextuple algebra which may be called (bd_6), of which the multiplication table is †

(bd_6)	i	j	k	l	m	n
i	i	j	k	l	m	n
j	j	k	0	0	0	m
k	k	0	0	0	0	0
l	l	ak	0	k	0	bm
m	0	0	0	0	0	0
n	0	0	0	0	0	0

[132^2]. The defining equation of this case is

$$jn = 0,$$

and there is no pure algebra in this case.

[13^2]. The defining equations of this case are

$$j^2 = k, \quad lj = k, \quad jk = jl = kj = k^2 = kl = lk = l^2 = 0,$$

which give

$$. \, km = kn = 0.$$

There is a sextuple algebra in this case which may be called (be_6), of which the multiplication table is *

(be_6)	i	j	k	l	m	n
i	i	j	k	l	m	n
j	j	k	0	0	0	am
k	k	0	0	0	0	0
l	l	k	0	0	0	bm
m	0	0	0	0	0	0
n	0	0	0	0	0	0

[134]. The defining equations of this case are

$$jk = -kj = l, \quad j^2 = k^2 = jk = kj = kl = lk = l^2 = 0.$$

There is a sextuple algebra in this case which may be called (bf_6), of which the multiplication table is *

(bf_6)	i	j	k	l	m	n
i	i	j	k	l	m	n
j	j	0	0	k	0	m
k	k	0	0	0	0	0
l	l	$-k$	0	0	0	am
m	0	0	0	0	0	0
n	0	0	0	0	0	0

---

* In relative form, $i = A:A + B:B + C:C + D:D$, $j = A:B - C:D$, $k = A:D$, $l = A:C + B:D$, $m = A:E$, $n = B:E + aC:E$. [C. S. P.]

[14]. The defining equations of this case are

$$ij = ji = j, \quad ik = ki = k, \quad il = li = l, \quad im = m, \quad ni = n, \quad mi = in = 0,$$

which give

$$0 = jm = jn = km = kn = lm = ln = mj = nj = mk = nk = ml = nl = m^2 = nm = n^2.$$

There are four cases defined as in [13].

[141]. The defining equations of this case are

$$j^2 = k, \quad jk = kj = l, \quad jl = k^2 = kl = lj = lk = l^2 = 0,$$

which give

$$mn = d_{56}l.$$

There is a sextuple algebra which may be called (bg_6), of which the multiplication table is*

(bg_6)	i	j	k	l	m	n
i	i	j	k	l	m	0
j	j	k	l	0	0	0
k	k	l	0	0	0	0
l	l	0	0	0	0	0
m	0	0	0	0	0	l
n	n	0	0	0	0	0

[142]. The defining equations of this case are the same as in [132], which give

$$mn = c_{56}k.$$

---

* In relative form, $i = A:A + B:B + C:C + D:D$, $j = A:B + B:C + C:D$, $k = A:C + B:D$, $l = A:D$, $m = A:E$, $n = E:D$. [C. S. P.]

There is a sextuple algebra which may be called (bh_6), of which the multiplication table is *

(bh_6)	i	j	k	l	m	n
i	i	j	k	l	m	0
j	j	k	0	0	0	0
k	k	0	0	0	0	0
l	l	ak	0	k	0	0
m	0	0	0	0	0	k
n	n	0	0	0	0	0

[143]. The defining equations of this case are the same as in [13^2]. There is a sextuple algebra which may be called (bi_6), of which the multiplication table is †

(bi_6)	i	j	k	l	m	n
i	i	j	k	l	m	0
j	j	k	0	0	0	0
k	k	0	0	0	0	0
l	l	k	0	0	0	0
m	0	0	0	0	0	k
n	n	0	0	0	0	0

---

* This algebra has two varieties, analogous to those of (c_3). The first is, in relative form, $i = A : A + B : B + C : C + D : D$, $j = A : B + B : C + A : D$, $k = A : C$, $l = -A : B + D : C$, $m = A : E$, $n = E : C$. The second in relative form is the same except that $j = A : B + b^{-1}D : C$, $l = A : D - bB : C$. [C. S. P.]

† This algebra may be slightly simplified by putting $j - l$ for j. Then, in relative form, $i = A : A + B : B + C : C$, $j = B : C$, $k = A : C$, $l = A : B$, $m = A : D$, $n = D : C$. [C. S. P.]

$[14^2]$. The defining equations of this case are the same as in [134]. There is a sextuple algebra which may be called (bj_6), of which the multiplication table is *

(bj_6)	i	j	k	l	m	n
i	i	j	k	l	m	0
j	j	0	0	k	0	0
k	k	0	0	0	0	0
l	l	$-k$	0	0	0	0
m	0	0	0	0	0	k
n	n	0	0	0	0	0

[15]. The defining equations of this case are

$$ij = ji = j, \quad ik = ki = k, \quad il = l, \quad im = m, \quad in = n, \quad li = mi = ni = 0,$$

which give

$$j^2 = k, \quad 0 = jk = kj = k^2 = lj = lk = l^2 = lm = ln = mj = mk = ml = m^2$$
$$= mn = nj = nk = nl = nm = n^2.$$

There is a sextuple algebra which may be called (bk_6), of which the multiplication table is †

(bk_6)	i	j	k	l	m	n
i	i	j	k	l	m	n
j	j	k	0	m	n	0
k	k	0	0	n	0	0
l	0	0	0	0	0	0
m	0	0	0	0	0	0
n	0	0	0	0	0	0

[16]. The defining equations of this case are

$$ij = ji = j, \quad ik = ki = k, \quad il = l, \quad im = m, \quad ni = n, \quad li = mi = in = 0,$$

which give

$$j^2 = k, \quad 0 = jk = jn = kj = k^2 = kn = lj = lk = l^2 = lm = mj = mk = nl = m^2$$
$$= nj = nk = nl = nm = n^2.$$

There is a sextuple algebra which may be called (bl_6), of which the multiplication table is *

(bl_6)	i	j	k	l	m	n
i	i	j	k	l	m	0
j	j	k	0	m	0	0
k	k	0	0	0	0	0
l	0	0	0	0	0	k
m	0	0	0	0	0	0
n	n	0	0	0	0	0

---

* In relative form, $i = A:A + B:B + C:C$, $j = A:B + B:C$, $k = A:C$, $l = B:D + A:E$, $m = A:D$, $n = E:C$. [C. S. P.]

[17]. The defining equations of this case are

$$ij = ji = j, \quad ik = ki = k, \quad il = l, \quad mi = m, \quad li = im = in = ni = 0.$$

There is no pure algebra in this case.

[18]. The defining equations of this case are

$$ij = ji = j, \quad ik = k, \ il = l, \quad im = m, \quad in = n, \quad ki = li = mi = ni = 0.$$

There is no pure algebra in this case.

[19]. The defining equations of this case are

$$ij = ji = j, \quad ik = k, \quad il = l, \quad im = m, \quad ni = n, \quad in = kl = li = ni = n.$$

There is no pure algebra in this case.

[10′]. The defining equations of this case are

$$ij = ji = j, \quad ik = k, \quad il = l, \quad mi = m, \quad ni = n, \quad im = in = ki = li = 0.$$

There is no pure algebra in this case.

[11′]. The defining equations of this case are

$$ij = ji = j, \quad ik = k, \quad il = l, \quad mi = m, \quad im = li = in = ni = 0.$$

There is no pure algebra in this case.

[12′]. The defining equations of this case are

$$ij = ji = j, \quad ik = k, \quad li = l, \quad il = im = in = ki = mi = ni = 0.$$

There is no pure algebra in this case.

[13′]. The defining equations of this case are

$$ij = j, \quad ik = k, \quad il = l, \quad im = m, \quad in = n, \quad ji = ki = li = mi = ni = 0.$$

There is no pure algebra in this case.

[14′]. The defining equations of this case are

$$ij = j, \quad ik = k, \quad il = l, \quad im = m, \quad ni = n, \quad ji = ki = li = mi = in = 0.$$

There is no pure algebra in this case.

[15′]. The defining equations of this case are

$$ij = j, \quad ik = k, \quad il = l, \quad mi = m, \quad ni = n, \quad im = in = ji = ki = li = 0.$$

There is no pure algebra in this case.

[16′]. The defining equations of this case are

$$ij = j, \quad ik = k, \quad il = l, \quad mi = m, \quad im = in = jk = kl = li = ni = 0.$$

There is no pure algebra in this case.

[17′]. The defining equations of this case are

$$ij = j, \quad ik = k, \quad li = l, \quad mi = m, \quad il = im = in = ji = ki = ni = 0.$$

There is no pure algebra in this case.

[18′]. The defining equations of this case are

$$ij = j, \quad ik = k, \quad li = l, \quad ji = ki = il = im = in = mi = ni = 0.$$

There are six cases :

[18′1], when $m^2 = m$, $mn = n$, $nm = 0$,

[18′2], when $m^2 = m$, $mn = 0$, $nm = n$,

[18′3], when $m^2 = n$, $mn = nm = 0$, $n^2 = m$,

[18′4], when $m^2 = m$, $mn = nm = n^2 = 0$,

[18′5], when $m^2 = n$, $m^3 = 0$,

[18′6], when $m^2 = n^2 = 0$.

[18′1]. The defining equations of this case are

$$m^2 = m, \quad mn = n, \quad nm = 0.$$

There are two cases :

[18′1²], when $ml = 0$;

[18′12], when $ml = l$.

[18′1²]. The defining equation of this case is

$$ml = 0.$$

There is no pure algebra in this case.

[18′12]. The defining equation of this case is

$$ml = l.$$

There are two cases :

[18′121], when $jm = j$;

[18′12²], when $jm = 0$.

[18′121]. The defining equation of this case is

$$jm = j.$$

There is a sextuple algebra which may be called (bm_6), of which the multiplication table is *

---

* In relative form, $i = A : A$, $j = A : B$, $k = A : C$, $l = B : A$, $m = B : B$, $n = B : C$. [C. S. P.]

(bm_6)	i	j	k	l	m	n
i	i	j	k	0	0	0
j	0	0	0	i	j	k
k	0	0	0	0	0	0
l	l	m	n	0	0	0
m	0	0	0	l	m	n
n	0	0	0	0	0	0

[$18'12^2$]. The defining equation of this case is

$$jm = 0.$$

There is no pure algebra in this case.

[$18'2$]. The defining equations of this case are

$$m^2 = m, \quad mn = 0, \quad nm = n.$$

There are two cases:

[$18'21$], when $ml = l$;
[$18'2^2$], when $ml = 0$.

[$18'21$]. The defining equation of this case is

$$ml = l.$$

There are two cases:

[$18'21^2$], when $jm = j$;
[$18'212$], when $jm = 0$.

[$18'21^2$]. The defining equation of this case is

$$jm = j.$$

There is no pure algebra in this case.

[$18'212$]. The defining equation of this case is

$$jm = 0.$$

There is no pure algebra in this case.

[18'2²]. The defining equation of this case is

$$ml = 0 .$$

There is no pure algebra in this case.

[18'3]. The defining equations of this case are

$$m^2 = m , \quad mn = nm = 0 , \quad n^2 = n .$$

There is no pure algebra in this case.

[18'4]. The defining equations of this case are

$$m^2 = m , \quad mn = nm = n^2 = 0 .$$

There are two cases :

[18'41], when $jm = j$;
[18'42], when $jm = 0$.

[18'41]. The defining equation of this case is

$$jm = j .$$

There is no pure algebra in this case.

[18'42]. The defining equation of this case is

$$jm = 0 .$$

There is no pure algebra in this case.

[18'5]. The defining equations of this case are

$$m^2 = n , \quad m^3 = 0 .$$

There is no pure algebra in this case.

[18'6]. The defining equations of this case are

$$m^2 = n^2 = 0 .$$

There is no pure algebra in this case.

[19']. The defining equations of this case are

$$ij = j , \quad ki = k , \quad ji = ik = il = im = in = li = mi = ni = 0 .$$

There is no pure algebra in this case.

[2]. The algebras belonging to this case are not investigated, because it is evident from § 69 that they are rarely of use unless combined with an idempotent basis, so as to give septuple algebras.

Natural Classification.

There are many cases of these algebras which may obviously be combined into natural classes, but the consideration of this portion of the subject will be reserved to subsequent researches.

ADDENDA.

I.

On the Uses and Transformations of Linear Algebra.

BY BENJAMIN PEIRCE.

[*Presented* to the American Academy of Arts and Sciences, May 11, 1875.]

Some definite interpretation of a linear algebra would, at first sight, appear indispensable to its successful application. But on the contrary, it is a singular fact, and one quite consonant with the principles of sound logic, that its first and general use is mostly to be expected from its want of significance. The interpretation is a trammel to the use. Symbols are essential to comprehensive argument. The familiar proposition that all A is B, and all B is C, and therefore all A is C, is contracted in its domain by the substitution of significant words for the symbolic letters. The A, B, and C, are subject to no limitation for the purposes and validity of the proposition; they may represent not merely the actual, but also the ideal, the impossible as well as the possible. In Algebra, likewise, the letters are symbols which, passed through a machinery of argument in accordance with given laws, are developed into symbolic results under the name of formulas. When the formulas admit of intelligible interpretation, they are accessions to knowledge; but independently of their interpretation they are invaluable as symbolical expressions of thought. But the most noted instance is the symbol called the impossible or imaginary, known also as the square root of minus one, and which, from a shadow of meaning attached to it, may be more definitely distinguished as the symbol of *semi-inversion*. This symbol is restricted to a precise signification as the representative of perpendicularity in quaternions, and this wonderful algebra of space is intimately dependent upon the special use of the symbol for its symmetry, elegance, and power. The immortal author of quaternions has shown that there are other significations which may attach to the symbol in other cases. But the strongest use of the symbol is to be found in its magical power of doubling the actual universe, and placing by its side an ideal universe, its exact counterpart, with which it can be compared and contrasted, and, by means of curiously connecting fibres, form with it an organic whole, from which modern analysis has developed her surpassing geometry. The letters or units of the linear algebras, or to use the better term proposed by Mr. Charles S. Peirce, the *vids* of these algebras, are fitted to perform a similar function each in its peculiar way. This is their primitive and perhaps will always be their principal use. It does not exclude the possibility of some special modes of interpretation, but, on the contrary, a higher philosophy, which believes in the capacity of the material universe for all expressions of human thought, will find, in the utility of the vids, an indication of their probable reality of interpretation. Doctor Hermann Hankel's alternate numbers, with Professor Clifford's applications to determinants, are a curious and interesting example of the possible advantage to be obtained from the new algebras. Doctor Spottiswoode in his fine, generous, and complete analysis of my own treatise before the London Mathematical Society in November of 1872, has regarded these numbers as quite different from the algebras discussed in my treatise, because they are neither linear nor limited. But there is no difficulty in reducing them to a linear form, and, indeed, my algebra (e_3) is the simplest case of Hankel's alternate numbers; and in any other case, in which n is the number of the Hankel elements employed, the complete number of vids of the corresponding linear algebra is $2^n - 1$. The limited character of the algebras which I have investigated may be regarded as an accident of the mode of discussion. There is, however, a large number of unlimited algebras suggested by the investigations, and Hankel's numbers themselves would have been a natural generalization from the proposition of § 65 of my algebra.[*] Another class of unlimited algebras, which would readily occur from the inspection of those which are given, is that in which all the powers of a vid are adopted as independent vids, and the highest power may either be zero, or unity, or the vid itself, and the zero power of the fundamental vid, *i. e.* unity itself, may also be retained as a vid. But I desire to draw especial attention to that class, which is also unlimited, and for which, when it was laid before the mathematical society of London in January of 1870, Professor Clifford proposed the appropriate name of *quadrates.*

---

[*] This remark is not intended as a foundation for a claim upon the Hankel numbers, which were published in 1867, three years prior to the publication of my own treatise.—B. P. [They were given much earlier under the name of *clefs* by Cauchy, and (substantially) at a still earlier date by Grassmann. —C. S. P.]

Quadrates.

The best definition of quadrates is that proposed by Mr. Charles S. Peirce. If the letters A, B, C, etc., represent absolute quantities, differing in quality, the vids may represent the relations of these quantities, and may be written in the form

$$(A:A)\,(A:B)\,(A:C) \;\ldots\; (B:A)\,(B:B) \;\ldots\; (C:A), \text{ etc.}$$

subject to the equations

$$(A:B)\,(B:C) = (A:C)$$
$$(A:B)\,(C:D) = 0.$$

In other words, every product vanishes, in which the second letter of the multiplier differs from the first letter of the multiplicand; and when these two letters are identical, both are omitted, and the product is the vid which is compounded of the remaining letters, which retain their relative position.

Mr. Peirce has shown by a simple logical argument that the quadrate is the legitimate form of a complete lineaı algebra. and that all the forms of the algebras given by me must be imperfect quadrates, and has confirmed this conclusion by actual investigation and reduction. His investigations do not however dispense with the analysis by which the independent forms have been deduced in my treatise, though they seem to throw much light upon their probable use.

Unity.

The sum of the vids $(\overset{.}{A}:A)$, $(B:B)$, $(C:C)$, etc., extended so as to include all the letters which represent absolute quantities in a given algebra, whether it be a complete or an incomplete quadrate has the peculiar character of being idempotent, and of leaving any factor unchanged with which it is combined as multiplier or multiplicand. This is the distinguishing property of unity, so that this combination of the vids can be regarded as unity, and may be introduced as such and called the *vid of unity*. There is no other combination which possesses this property.

But any one of the vids $(A:A)$, $(B:B)$, etc., or the sum of any of these vids is idempotent. There are many other idempotent combinations, such as

$$(A:A) + x\,(A:B), \quad y\,(A:B) + (B:B),$$
$$\tfrac{1}{2}\,(A:A) + \tfrac{1}{2}\,(A:B) + \tfrac{1}{2}\,(B:A) + \tfrac{1}{2}\,(B:B),$$

which may deserve consideration in making transformations of an algebra preparatory to its application.

Inversion.

A vid which differs from unity, but of which the square is equal to unity, may be called a *vid of inversion.* For such a vid when applied to some other combination transforms it; but, whatever the transformation, a repetition of the application restores the combination to its primitive form. A very general form of a vid of inversion is

$$(A : A) \pm (B : B) \pm (C : C) \pm \text{etc.,}$$

in which each doubtful sign corresponds to two cases, except that at least one of the signs must be negative. The negative of unity might also be regarded as a symbol of inversion, but cannot take the place of an independent vid. Besides the above vids of inversion, others may be formed by adding to either of them a vid consisting of two different letters, which correspond to two of the one-lettered vids of different signs; and this additional vid may have any numerical coefficient whatever. Thus

$$(A : A) + (B : B) - (C : C) + x (A : C) + y (B : C)$$

is a vid of inversion.

The new vid which Professor Clifford has introduced into his biquaternions is a vid of inversion.

Semi-Inversion.

A vid of which the square is a vid of inversion, is a *vid of semi-inversion.* A very general form of a vid of semi-inversion is

$$(A : A) \pm (B : B) \pm \mathsf{J}(C : C) \pm \text{etc.}$$

in which one or more of the terms $(A : A)$, $(B : B)$, etc., have J for a coefficient. The combination

$$(A : A) \pm \mathsf{J}(B : B) + x(A : B) + \text{etc.}$$

is also a vid of semi-inversion. With the exception of unity, all the vids of Hamilton's quaternions are vids of semi-inversion.

The Use of Commutative Algebras.

Commutative algebras are especially applicable to the integration of differential equations of the first degree with constant coefficients. If i, j, k, etc., are the vids of such an algebra, while x, y, z, etc., are independent variables, it is easy to show that a solution may have the form $F(xi + yj + zk$ $+$ etc.), in which F is an arbitrary function, and i, j, k, etc., are connected by some simple equation. This solution can be developed into the form

$$F(xi + yj + zk + \text{etc.}) = Mi + Nj + Pk + \text{etc.}$$

in which M, N, P, etc., will be functions of x, y, z, etc., and each of them is a solution of the given equation. Thus in the case of Laplace's equation for the potential of attracting masses, the vids must satisfy the equation

$$i^2 + j^2 + k^2 = 0 .$$

The algebra (a_3) of which the multiplication table is

	i	j	k
i	i	j	k
j	j	k	0
k	k	0	0

may be used for this case. Combinations i_1, j_1, k_1 of these vids can be found which satisfy the equation

$$i_1^2 + j_1^2 + k_1^2 = 0 ,$$

and if the functional solution

$$F(xi_1 + yj_1 + zk_1)$$

is developed into the form of the original vids

$$Mi + Nj + Pk,$$

M, N, and P will be independent solutions, of such a kind that the surfaces for which N and P are constant will be perpendicular to that for which M is constant, which is of great importance in the problems of electricity.

The Use of Mixed Algebras.

It is quite important to know the various kinds of pure algebra in making a selection for special use, but mixed algebras can also be used with advantage in certain cases. Thus, in Professor Clifford's biquaternions, of which he has demonstrated the great value, other vids can be substituted for unity and his new vid, namely their half sum and half difference, and each of the original vids of the quaternions can be multiplied by these, giving us two sets of vids, each of which will constitute an independent quadruple algebra of the same form with quaternions. Thus if i, j, k, are the primitive quaternion vids and w the new vid, let

$$a_1 = \tfrac{1}{2}(1 + w). \qquad\qquad a_2 = \tfrac{1}{2}(1 - w).$$
$$i_1 = a_1 i. \qquad\qquad\qquad i_2 = a_2 i.$$
$$j_1 = a_1 j. \qquad\qquad\qquad j_2 = a_2 j.$$
$$k_1 = a_1 k. \qquad\qquad\qquad k_2 = a_2 k.$$

Then since

$$a_1^2 = a_1. \qquad\qquad\qquad a_2^2 = a_2.$$
$$i_1^2 = j_1^2 = k_1^2 = -a_1. \qquad i_2^2 = j_2^2 = k_2^2 = -a_1.$$
$$i_1 j_1 = k_1 = -j_1 i_1. \qquad i_2 j_2 = k_2 = -j_2 i_2.$$
$$j_1 k_1 = i_1 = -k_1 j_1. \qquad j_2 k_2 = i_2 = -k_2 j_2.$$
$$k_1 i_1 = j_1 = -i_1 k_1. \qquad k_2 i_2 = j_2 = -i_2 k_2.$$
$$a_1 a_2 = 0 = a_2 a_1.$$
$$M_1 N_2 = 0 = N_2 M_1.$$

in which M_1 denotes any combination of the vids of the first algebra, and N_2 any combination of those of the second algebra. It may perhaps be claimed that these algebras are not independent, because the sum of the vids a_1 and a_2 is absolute unity. This, however, should be regarded as a fact of interpretation which is not apparent in the defining equations of the algebras.

II.

On the Relative Forms of the Algebras.

BY C. S. PEIRCE.

Given an associative algebra whose letters are i, j, k, l, etc., and whose multiplication table is

$$i^2 = a_{11} i + b_{11} j + c_{11} k + \text{etc.*}$$
$$ij = a_{12} i + b_{12} j + c_{12} k + \text{etc.}$$
$$ji = a_{21} i + b_{21} j + c_{21} k + \text{etc.},$$
$$\text{etc., etc.}$$

I proceed to explain what I call the relative form of this algebra.

---

*I have used a_{11}, etc., in place of the a_1, etc., used by my father in his text.

Let us assume a number of new units, A, I, J, K, L, etc., one more in number than the letters of the algebra, and every one except the first, A, corresponding to a particular letter of the algebra. These new units are susceptible of being multiplied by numerical coefficients and of being added together;[*] but they cannot be multiplied together, and hence are called *non-relative* units.

Next, let us assume a number of operations each denoted by bracketing together two non-relative units separated by a colon. These operations, equal in number to the square of the number of non-relative units, may be arranged as follows:

$$(A:A) \quad (A:I) \quad (A:J) \quad (A:K), \text{ etc.}$$
$$(I:A) \quad (I:I) \quad (I:J) \quad (I:K), \text{ etc.}$$
$$(J:A) \quad (J:I) \quad (J:J) \quad (J:K), \text{ etc.}$$

Any one of these operations performed upon a polynomial in non-relative units, of which one term is a numerical multiple of the letter following the colon, gives the same multiple of the letter preceding the colon. Thus, $(I:J)(aI + bJ + cK) = bI.$[†] These operations are also taken to be susceptible of associative combination. Hence $(I:J)(J:K) = (I:K)$; for $(J:K) K = J$ and $(I:J) J = I$, so that $(I:J)(J:K)K = I$. And $(I:J)(K:L) = 0$; for $(K:L) L = K$ and $(I:J) K = (I:J)(0.J + K) = 0.I = 0$. We further assume the application of the distributive principle to these operations; so that, for example,

$$\{(I:J) + (K:J) + (K:L)\}(aJ + bL) = aJ + (a + b) K.$$

Finally, let us assume a number of complex operations denoted by i', j', k', l', etc., corresponding to the letters of the algebra and determined by its multiplication table in the following manner:

$$i' = (I:A) + a_{11}(I:I) + b_{11}(J:I) + c_{11}(K:I) + \text{ etc.}$$
$$+ a_{12}(I:J) + b_{12}(J:J) + c_{12}(K:J) + \text{ etc.}$$
$$+ a_{13}(I:K) + b_{13}(J:K) + c_{13}(K:K) + \text{ etc.} + \text{etc.}$$
$$j' = (J:A) + a_{21}(I:I) + b_{21}(J:I) + c_{21}(K:I) + \text{ etc.}$$
$$+ a_{22}(I:J) + b_{22}(J:J) + c_{22}(K:J) + \text{ etc.}$$
$$+ a_{23}(I:K) + b_{23}(J:K) + c_{23}(K:K) + \text{ etc.} + \text{ etc.}$$
$$k' = \text{ etc.}$$

---

[*] Any one of them multiplied by 0 gives 0.   [†] If $b = 0$, of course the result is 0.

Any two operations are equal which, being performed on the same operand, invariably give the same result. The ultimate operands in this case are the non-relative units. But any operations compounded by addition or multiplication of the operations i', j', k', etc., if they give the same result when performed upon A, will give the same result when performed upon any one of the non-relative units. For suppose $i'j'A = k'l'A$. We have

$$i'j'A = i'J = a_{12}I + b_{12}J + c_{12}K + \text{etc.}$$
$$k'l'A = k'L = a_{34}I + b_{34}J + c_{34}K + \text{etc.}$$

so that $a_{12} = a_{34}$, $b_{12} = b_{34}$, $c_{12} = c_{34}$, etc., and in our original algebra $ij = kl$. Hence, multiplying both sides of the equation into any letter, say m, $ijm = klm$. But

$$ijm = i\,(a_{25}i + b_{25}j + c_{25}k + \text{etc.}) = (a_{11}a_{25} + a_{12}b_{25} + a_{13}c_{25} + \text{etc.})\,i$$
$$+ (b_{11}a_{25} + b_{12}b_{25} + b_{13}c_{25} + \text{etc.})\,j + \text{etc.}$$

But we have equally

$$i'j'm'A = (a_{11}a_{25} + a_{12}b_{25} + a_{13}c_{25} + \text{etc.})\,I + (b_{11}a_{25} + b_{12}b_{25} + b_{13}c_{25} + \text{etc.})\,J + \text{etc.}$$

So that $i'j'm'A = k'l'm'A$. Hence, $i'j'M = k'l'M$. It follows, then, that if $i'j'A = k'l'A$, then $i'j'$ into any non-relative unit equals $k'l'$ into the same unit, so that $i'j' = k'l'$. We thus see that whatever equality subsists between compounds of the accented letters i', j', k', etc., subsists between the same compounds of the corresponding unaccented letters i, j, k, so that the multiplication tables of the two algebras are the same.* Thus, what has been proved is that any associative algebra can be put into relative form, *i. e.* (see my *brochure* entitled *A brief Description of the Algebra of Relatives*) that every such algebra may be represented by a matrix.

Take, for example, the algebra (bd_5). It takes the relative form

$i = (I:A) + (J:I) + (L:K), \quad j = (J:A),$
$k = (K:A) + (J:I) + \mathfrak{r}(L:I) + (I:K) + (M:K) + \mathfrak{r}(J:L) - (J:M) - \mathfrak{r}(L:M),$
$l = (L:A) + (J:K), \quad m = (M \cdot A) + (\mathfrak{r}^2 - 1)(J:I) - (L \cdot K) \quad \mathfrak{r}^2(J:M).$

---

* A brief proof of this theorem, perhaps essentially the same as the above, was published by me in the *Proceedings of the American Academy of Arts and Sciences*, for May 11, 1875.

This is the same as to say that the general expression $xi + yj + zk + ul + vm$ of this algebra has the same laws of multiplication as the matrix

$$
\begin{array}{cccccc}
0, & 0, & 0, & 0, & 0, & 0, \\
x, & 0, & 0, & z, & 0, & 0, \\
y, & \begin{array}{c} x+z \\ +(\mathfrak{r}^2-1)v, \end{array} & 0, & u, & \mathfrak{r}z, & -z-\mathfrak{r}^2v, \\
z, & 0, & 0, & 0. & 0, & 0, \\
u, & \mathfrak{r}z, & 0, & x-v, & 0, & -\mathfrak{r}z \\
v, & 0, & 0, & z, & 0, & 0.
\end{array}
$$

Of course, every algebra may be put into relative form in an infinity of ways; and simpler ways than that which the rule affords can often be found. Thus, for the above algebra, the form given in the foot-note is simpler, and so is the following:

$$i = (B:A) + (C:B) + (F:D) + (C:E), \quad j = (C:A),$$
$$k = (D:A) + (E:D) + (C:B) + \mathfrak{r}(F:B) + \mathfrak{r}(C:F),$$
$$l = (F:A) + (C:D), \quad m = (E:A) + (\mathfrak{r}^2-1)(C:B) - (B:A) - (F:D) - (C:E).$$

These different forms will suggest transformations of the algebra. Thus, the relative form in the foot-note to (bd_5) suggests putting

$$i_1 = i + m, \quad j_1 = \mathfrak{r}^2 j, \quad k_1 = k + \mathfrak{r}^{-1}i + \mathfrak{r}^{-1}m, \quad l_1 = \mathfrak{r}l + j, \quad m_1 = -m,$$

when we get the following multiplication table, where ρ is put for $\mathfrak{r}^{-1}$:

	i	j	k	l	m
i	0	0	0	0	j
j	0	0	0	0	0
k	0	0	i	j	l
l	0	0	ρj	0	0
m	$\rho^2 j$	0	ρl	0	j

Ordinary algebra with imaginaries, considered as a double algebra, is, in relative form,

$$1 = (X:X) + (Y:Y), \quad \mathsf{J} = (X:Y) - (Y:X).$$

This shows how the operation 'J turns a vector through a right angle in the plane of X, Y. Quaternions in relative form is

$$1 = (W:W) + (X:X) + (Y:Y) + (Z:Z),$$
$$i = (X:W) - (W:X) + (Z:Y) - (Y:Z),$$
$$j = (Y:W) - (Z:X) - (W:Y) + (X:Z),$$
$$k = (Z:W) + (Y:X) - (X:Y) - (W:Z).$$

We see that we have here a reference to a space of four dimensions corresponding to X, Y, Z, W.

III.

On the Algebras in which Division is Unambiguous.

BY C. S. PEIRCE.

1. In the *Linear Associative Algebra*, the coefficients are permitted to be imaginary. In this note they are restricted to being real. It is assumed that we have to deal with an algebra such that from $AB = AC$ we can infer that $A = 0$ or $B = C$. It is required to find what forms such an algebra may take.

2. If $AB = 0$, then either $A = 0$ or $B = 0$. For if not, $AC = A(B + C)$, although A does not vanish and C is unequal to $B + C$.

3. The reasoning of § 40 holds, although the coefficients are restricted to being real. It is true, then, that since there is no expression (in the algebra under consideration) whose square vanishes, there must be an expression, i, such that $i^2 = i$.

4. By § 41, it appears that for every expression in the algebra we have

$$iA = Ai = A.$$

5. By the reasoning of §53, it appears that for every expression A there is an equation of the form

$$\Sigma_m (a_m A^m) + bi = 0.$$

But i is virtually arithmetical unity, since $iA = Ai = A$; and this equation may be treated by the ordinary theory of equations. Suppose it has a real root, α; then it will be divisible by $(A - \alpha)$, and calling the quotient B we shall have

$$(A - \alpha i) B = 0.$$

But $A - \alpha i$ is not zero, for A was supposed dissimilar to i. Hence a product of finites vanishes, which is impossible. Hence the equation cannot have a real root. But the whole equation can be resolved into quadratic factors, and some one of these must vanish. Let the irresoluble vanishing factor be

$$(A - s)^2 + t^2 = 0.$$

Then

$$\left(\frac{A - s}{t}\right)^2 = -1,$$

or, every expression, upon subtraction of a real number (*i. e.* a real multiple of i), can be converted, in one way only, into a quantity whose square is a negative number. We may express this by saying that every quantity consists of a scalar and a vector part. A quantity whose square is a negative number we here call a *vector*.

6. Our next step is to show that the vector part of the product of two vectors is linearly independent of these vectors and of unity. That is, i and j being any two vectors,* if

$$ij = s + v$$

where s is a scalar and v a vector, we cannot determine three real scalars a, b, c, such that

$$v = a + bi + cj.$$

This is proved, if we prove that no scalar subtracted from ij leaves a remainder $bi + cj$. If this be true when i and j are any unit vectors whatever, it is true when these are multiplied by real scalars, and so is true of every pair of vectors. We will, then, suppose i and j to be unit vectors. Now,

$$ij^2 = -i.$$

If therefore we had

$$ij = a + bi + cj,$$

we should have

$$-i = ij^2 = aj + bij - c = ab - c + b^2i + (a + bc)j;$$

whence, i and j being dissimilar,

$$-i = b^2i, \qquad b^2 = -1,$$

and b could not be real.

---

* The idempotent basis having been shown to be arithmetical unity. we are free to use the letter i to denote another unit.

7. Our next step is to show that, i and j being any two vectors, and

$$ij = s + v,$$

s being a scalar and v a vector, we have

$$ji = r(s - v),$$

where r is a real scalar. It will be obviously sufficient to prove this for the case in which i and j are unit vectors. Assuming them such, let us write

$$ji = s' + v', \qquad vv' = s'' + v'',$$

where s' and s'' are scalars, while v' and v'' are vectors. Then

$$ij \cdot ji = (s + v)(s' + v') = ss' + sv' + s'v + v'' + s''.$$

But we have

$$ij \cdot ji = ij^2 i = -i^2 = 1.$$

Hence,

$$v'' = 1 - ss' - s'' - sv' - s'v.$$

But v'' is the vector of vv', so that by the last paragraph such an equation cannot subsist unless v'' vanishes. Thus we get

$$0 = 1 - ss' - s'' - sv' - s'v,$$

or

$$sv' = 1 - ss' - s'' - s'v.$$

But a quantity can only be separated in one way into a scalar and a vector part ; so that

$$sv' = - s'v.$$

That is,

$$ji = \frac{s'}{s}(s - v). \quad Q.E.D.$$

8. Our next step is to prove that $s = s'$; so that if $ij = s + v$ then $ji = s - v$. It is obviously sufficient to prove this when i and j are unit vectors. Now from any quantity a scalar may be subtracted so as to leave a remainder whose square is a scalar. We do not yet know whether the sum of two vectors is a vector or not (though we do know that it is not a scalar). Let us then take such a sum as $ai + bj$ and suppose x to be the scalar which subtracted from it makes the square of the remainder a scalar. Then, C being a scalar,

$$(-x + ai + bj)^2 = C.$$

But developing the square we have

$$(-x + ai + bj)^2 = x^2 - a^2 - b^2 + abs + abs' - 2axi + 2bxj + ab\left(1 - \frac{s'}{s}\right)v = C;$$

i. e.

$$ab\left(1 - \frac{s'}{s}\right)v = C - x^2 + a^2 + b^2 - abs - abs' + 2axi + 2bxj.$$

But v being the vector of ij, by the last paragraph but one the equation must vanish. Either then $v = 0$ or $1 - \frac{s'}{s} = 0$. But if $v = 0$, $ij = s$, and multiplying into j,

$$-i = sj,$$

which is absurd, i and j being dissimilar. Hence $1 - \frac{s'}{s} = 0$ and

$$ji = s - v. \quad Q.E.D.$$

9. The number of independent vectors in the algebra cannot be two. For the vector of ij is independent of i and j. There may be no vector, and in that case we have the ordinary algebra of reals; or there may be only one vector, and in that case we have the ordinary algebra of imaginaries.

Let i and j be two independent vectors such that

$$ij = s + v.$$

Let us substitute for j

$$j_1 = si + j.$$

Then we have

$$ij_1 = v, \quad j_1 i = -v,$$
$$j_1 v = j_1 ij_1 = -j_1^2 i = i, \quad vj_1 = ij_1^2 = -i,$$
$$iv = i^2 j_1 = -j_1, \quad vi = ij_1 i = -j_1 i^2 = j_1.$$

Thus we have the algebra of real *quaternions*. Suppose we have a fourth unit vector, k, linearly independent of all the others, and let us write

$$j_1 k = s' + v',$$
$$ki = s'' + v''.$$

Let us substitute for k

$$k_1 = s'' i + s' j_1 + k,$$

and we get

$$j_1 k_1 = -s'' v + v', \quad k_1 j_1 = s'' v - v',$$
$$k_1 i = -s' v + v'', \quad ik_1 = s' v - v''.$$

Let us further suppose
$$(ij_1) k_1 = s' + v'''.$$
Then, because ij_1 is a vector,
$$k_1 (ij_1) = s''' - v'''.$$
But
$$k_1 j_1 = -j_1 k_1, \quad k_1 i = -ik_1,$$
because both products are vectors.

Hence
$$i . j_1 k_1 = -i . k_1 j_1 = -ik_1 . j_1 = k_1 i . j_1 = k_1 . ij_1.$$
Hence
$$s''' + v''' = s''' - v'''$$

or $v''' = 0$, and the product of the two unit vectors is a scalar. These vectors cannot, then, be independent, or k cannot be independent of $ij = v$. Thus it is proved that a fourth independent vector is impossible, and that ordinary real algebra, ordinary algebra with imaginaries, and real quaternions are the only associative algebras in which division by finites always yields an unambiguous quotient.

THREE CENTURIES
OF
SCIENCE IN AMERICA

An Arno Press Collection

Adams, John Quincy. **Report of the Secretary of State upon Weights and Measures.** 1821.

Archibald, Raymond Clare. **A Semicentennial History of the American Mathematical Society: 1888-1938** *and* **Semicentennial Addresses of the American Mathematical Society.** 2 vols. 1938.

Bond, William Cranch. **History and Description of the Astronomical Observatory of Harvard College** *and* **Results of Astronomical Observations Made at the Observatory of Harvard College.** 1856.

Bowditch, Henry Pickering. **The Life and Writings of Henry Pickering Bowditch.** 2 vols. 1980.

Bridgman, Percy Williams. **The Logic of Modern Physics.** 1927.

Bridgman, Percy Williams. **Philosophical Writings of Percy Williams Bridgman.** 1980.

Bridgman, Percy Williams. **Reflections of a Physicist.** 1955.

Bush, Vannevar. **Science the Endless Frontier.** 1955.

Cajori, Florian. **The Chequered Career of Ferdinand Rudolph Hassler.** 1929.

Cohen, I. Bernard, editor. **The Career of William Beaumont and the Reception of His Discovery.** 1980.

Cohen, I. Bernard, editor. **Benjamin Peirce: "Father of Pure Mathematics"** **in America.** 1980.

Cohen, I. Bernard, editor. **Aspects of Astronomy in America in the Nineteenth Century.** 1980.

Cohen, I. Bernard, editor. **Cotton Mather and American Science and Medicine: With Studies and Documents Concerning the Introduction of Inoculation or Variolation.** 2 vols. 1980.

Cohen, I. Bernard, editor. **The Life and Scientific Work of Othniel Charles Marsh.** 1980.

Cohen, I. Bernard, editor. **The Life and the Scientific and Medical Career of Benjamin Waterhouse: With Some Account of the Introduction of Vaccination in America.** 2 vols. 1980.

Cohen, I. Bernard, editor. **Research and Technology.** 1980.

Cohen, I. Bernard, editor. **Thomas Jefferson and the Sciences.** 1980.

Cooper, Thomas. **Introductory Lecture** *and* **A Discourse on the Connexion Between Chemistry and Medicine.** 2 vols. in one. 1812/1818.

Dalton, John Call. **John Call Dalton on Experimental Method.** 1980.

Darton, Nelson Horatio. **Catalogue and Index of Contributions to North American Geology: 1732-1891.** 1896.

Donnan, F[rederick] G[eorge] and Arthur Haas, editors. **A Commentary on the Scientific Writings of J. Willard Gibbs** *and* Duhem, Pierre. **Josiah-Willard Gibbs: A Propos de la Publication de ses Mémoires Scientifiques.** 3 vols. in two. 1936/1908.

Dupree, A[nderson] Hunter. **Science in the Federal Government: A History of Policies and Activities to 1940.** 1957.

Ellicott, Andrew. **The Journal of Andrew Ellicott.** 1803.

Fulton, John F. **Harvey Cushing: A Biography.** 1946.

Getman, Frederick H. **The Life of Ira Remsen.** 1940.

Goode, George Brown. **The Smithsonian Institution 1846-1896: The History of its First Half Century.** 1897.

Hale, George Ellery. **National Academies and the Progress of Research.** 1915.

Harding, T. Swann. **Two Blades of Grass: A History of Scientific Development in the U.S. Department of Agriculture.** 1947.

Hindle, Brooke. **David Rittenhouse.** 1964.

Hindle, Brooke, editor. **The Scientific Writings of David Rittenhouse.** 1980.

Holden, Edward S[ingleton]. **Memorials of William Cranch Bond, Director of the Harvard College Observatory, 1840-1859, and of his Son, George Phillips Bond, Director of the Harvard College Observatory, 1859-1865.** 1897.

Howard, L[eland] O[sslan]. **Fighting the Insects: The Story of an Entomologist, Telling the Life and Experiences of the Writer.** 1933.

Jaffe, Bernard. **Men of Science in America.** 1958.

Karpinski, Louis C. **Bibliography of Mathematical Works Printed in America through 1850.** Reprinted with **Supplement and Second Supplement.** 1940/1945.

Loomis, Elias. **The Recent Progress of Astronomy: Especially in the United States.** 1851.

Merrill, Elmer D. **Index Rafinesquianus: The Plant Names Published by C.S. Rafinesque with Reductions, and a Consideration of his Methods, Objectives, and Attainments.** 1949.

Millikan, Robert A[ndrews]. **The Autobiography of Robert A. Millikan.** 1950.

Mitchel, O[rmsby] M[acKnight]. **The Planetary and Stellar Worlds: A Popular Exposition of the Great Discoveries and Theories of Modern Astronomy.** 1848.

Organisation for Economic Co-operation and Development. **Reviews of National Science Policy: United States.** 1968.

Packard, Alpheus S. **Lamarck: The Founder of Evolution; His Life and Work.** 1901.

Pupin, Michael. **From Immigrant to Inventor.** 1930.

Rhees, William J. **An Account of the Smithsonian Institution.** 1859.

Rhees, William J. **The Smithsonian Institution: Documents Relative to its History.** 2 vols. 1901.

Rhees, William J. **William J. Rhees on James Smithson.** 2 vols. in one. 1980.

Scott, William Berryman. **Some Memories of a Palaeontologist.** 1939.

Shryock, Richard H. **American Medical Research Past and Present.** 1947.

Shute, Michael, editor. **The Scientific Work of John Winthrop.** 1980.

Silliman, Benjamin. **A Journal of Travels in England, Holland, and Scotland, and of Two Passages over the Atlantic in the Years 1805 and 1806.** 2 vols. 1812.

Silliman, Benjamin. **A Visit to Europe in 1851.** 2 vols. 1856

Silliman, Benjamin, Jr. **First Principles of Chemistry.** 1864.

Smith, David Eugene and Jekuthiel Ginsburg. **A History of Mathematics in America before 1900.** 1934.

Smith, Edgar Fahs. **James Cutbush: An American Chemist.** 1919.

Smith, Edgar Fahs. **James Woodhouse: A Pioneer in Chemistry, 1770-1809.** 1918.

Smith, Edgar Fahs. **The Life of Robert Hare: An American Chemist (1781-1858).** 1917.

Smith, Edgar Fahs. **Priestley in America: 1794-1804.** 1920.

Sopka, Katherine. **Quantum Physics in America: 1920-1935** (Doctoral Dissertation, Harvard University, 1976). 1980.

Steelman, John R[ay]. **Science and Public Policy: A Report to the President.** 1947.

Stewart, Irvin. **Organizing Scientific Research for War: The Administrative History of the Office of Scientifc Research and Development.** 1948.

Stigler, Stephen M., editor. **American Contributions to Mathematical Statistics in the Nineteenth Century.** 2 vols. 1980.

Trowbridge, John. **What is Electricity?** 1899.

True. Alfred. **Alfred True on Agricultural Experimentation and Research.** 1980.

True, F[rederick] W., editor. **The Semi-Centennial Anniversary of the National Academy of Sciences: 1863-1913** *and* **A History of the First Half-Century of the National Academy of Sciences: 1863-1913.** 2 vols. 1913.

Tyndall, John. **Lectures on Light: Delivered in the United States in 1872-73.** 1873.

U.S. House of Representatives. **Annual Report of the Board of Regents of the Smithsonian Institution...A Memorial of George Brown Goode together with a selection of his Papers on Museums and on the History of Science in America.** 1901.

U.S. National Resources Committee. **Research: A National Resource.** 3 vols. in one. 1938-1941.

U.S. Senate. **Testimony Before the Joint Commission to Consider the Present Organizations of the Signal Service, Geological Survey, Coast and Geodetic Survey, and the Hydrographic Office of the Navy Department.** 2 vols. 1866.